The Formation of
The Solar System
Theories Old and New

2nd Edition

The Formation of
The Solar System
Theories Old and New

2nd Edition

Michael Woolfson
University of York, UK

Summit Free Public Library

Imperial College Press

ICP

Published by

Imperial College Press
57 Shelton Street
Covent Garden
London WC2H 9HE

Distributed by

World Scientific Publishing Co. Pte. Ltd.

5 Toh Tuck Link, Singapore 596224

USA office: 27 Warren Street, Suite 401-402, Hackensack, NJ 07601

UK office: 57 Shelton Street, Covent Garden, London WC2H 9HE

Library of Congress Cataloging-in-Publication Data
Woolfson, Michael M. (Michael Mark), author.
    The formation of the solar system : theories old and new / Michael M. Woolfson, University of York, UK. -- 2nd edition.
       pages cm
    Includes bibliographical references and index.
    ISBN 978-1-78326-521-3 (hardcover : alk. paper) -- ISBN 978-1-78326-522-0 (pbk. : alk. paper)
    1. Solar system--Origin. I. Title.
    QB503.W659 2014
    523.2--dc23

       2014032549

British Library Cataloguing-in-Publication Data
A catalogue record for this book is available from the British Library.

Typeset by Stallion Press
Email: enquiries@stallionpress.com

Printed by FuIsland Offset Printing (S) Pte Ltd Singapore

Contents

Introduction

Most scientists think that the work they do is very important. Well, they would wouldn't they? It is a human trait, an aspect of vanity, to consider that what one does is more significant than it really is. You might think that scientists would be objective and self-critical and, to be fair, many of them are, but mostly they are prone to all the weaknesses of humanity at large. Nevertheless, if you were to ask a group of scientists which they thought was the most important, the most fundamental, of all scientific problems, the majority would probably reply that it is to understand the origin of the Universe. But, I hear you say, surely that it is a solved problem and, in one sense, it is. The big-bang theory starts with an event that occurs at an instant when there is no matter, no space and no time. A colossal release of energy at that 'beginning of everything' spreads out creating matter, space and time. There is some supporting evidence for this model. When we look at distant galaxies we find that they are receding from us at a speed proportional to their distance. If all motions were reversed then in 15 billion (thousand million) years they would all converge at the point where the Big Bang began — which can be regarded as anywhere since everywhere diverged from the same point at the beginning of time. Before 15 billion years ago there was no time — nothing existed of any kind.

I hope that you understand all that because I certainly do not. It is not that I do not *believe* in the big-bang theory — it is just that I do not really *understand* it. The mathematical model of the Big Bang is plain enough and many physicists and astronomers, including myself, can deal with that but I doubt that there are many people on this Earth that really *understand* it. My own test of whether or not I understand something is whether or not I can explain it to others. Sometimes in my teaching career, when I have been preparing a new course, I have suddenly realised that I could not provide a clear explanation for something — the reason being that the topic I *assumed* that I understood I did not really understand at all. Fortunately, in the teaching context, by reading and a bit of thought I have been able to deal with my own shortcomings. Nevertheless, I can promise you that I shall not be writing a book on the origin of the Universe.

So, if we now exclude the origin of the Universe and take it as an observational fact that we have matter, space and time, what would most scientists think of as the next most important problem? That surely must be to explain the origin of life. By what process can inanimate matter be transformed even into the most primitive life forms? Is it actually a spontaneous process? We are straying close to religious issues here so I shall go no further in that direction. But, once life exists, even in a primitive form, we do have a theory that *can* readily be understood, Darwin's theory of evolution, which leads us to the higher forms of life, including ourselves. Random genetic mutations occasionally create an individual that has advantages over others in its environment. The principle of 'survival of the fittest' ensures that its newly modified genes flourish and eventually become dominant and by small changes over long periods of time a new species can evolve. There is also some evidence for the occasional rather more rapid creation of new species. Something that Darwin's ideas do not deal with directly is the existence of consciousness, the knowledge of one's own identity and relationship to the outside world. That too is one of the difficult problems of science, although there are those who attempt to explain it in terms of computer technology by asserting that human beings resemble rather complex computers.

The topic of this book is not as important as those mentioned above and in many ways it is much more mundane. The Solar System is a collection of objects — the Sun, planets, satellites, etc. — made of ordinary matter — iron, silicates, ices and gases, the properties of which we well understand. The Universe is full of such material so all we have to do is to find a way of transforming it from one state to another. Having said that, although it is not an important problem in a fundamental sense, it is nevertheless quite an interesting one because it turns out to be very difficult. For more than two hundred years scientists have been struggling to find ways of just producing the Sun, planets and satellites, let alone all the other bodies of the Solar System. Part of the problem has been that scientists have tended to concentrate on parts of the system rather than looking at it as a whole. It is as though one was trying to understand the structure and workings of a car by studying just the wheels, the transmission system or the seats. It is only by looking at the whole car that one can understand the relationships of one part to another, how it works and how it was made. The same is true for the Solar System. It is not a collection of disconnected objects bearing little relationship to one another. It is a *system* and it will only be understood by examining it as a system.

The story that I tell goes back a long way — perhaps almost to the beginning of sentient mankind. Starting from the observations that a few points of light wandered around against the background of the stars, a picture emerged of a collection of bodies, connected to the Earth, Moon and Sun, that formed a separate family. Gradually the picture improved until, about three hundred years ago, we not only knew how all the bodies moved relative to each other but also understood the nature of the forces that made them move as they do.

Large telescopes, operating over a range of wavelengths from X-rays to radio waves, together with spacecraft, have given us detailed knowledge of virtually all the members of the solar-system family even to the extent of knowing something about the materials of which they consist. In addition, from Earth-bound observations we have even been able to detect planets around other stars — many of them — so we know that the Solar System is not a unique example of its kind. New knowledge has provided guidance for theoreticians attempting to explain the origin of the Solar System — but it also gives new constraints that their theories need to satisfy.

I began my study in this area, generally known as cosmogony, in 1962, intrigued by the fact that, while there had been many theories put forward, not one had survived close scientific scrutiny. While some of them were superficially attractive they all failed because they contravened some important scientific principle; it is a basic requirement of any theory that *every* aspect of it must be consistent with the science that we know. If a theory explains many things in the Solar System but is in conflict with scientific principles then it is wrong. You can no more have a nearly plausible theory than you can have a nearly pregnant woman. Armed with the knowledge that, since the

Solar System exists, there must be some viable theory for its origin I started on what turned out to be a long hard road. There were many dead ends and new beginnings. Ideas arose, seemed promising, failed critical tests and then were abandoned. However, one early basic idea, that was the core of what came to be called the Capture Theory, survived and evolved. What it evolved into is very different from the starting form but the essential idea is still there. Gradually a picture emerged that seemed to make sense and — a good sign — instead of problems piling up, as had been the original experience, it was solutions to problems that seemed to proliferate.

I have already confessed to my lack of deep understanding of the big-bang theory but I *do* understand the nature of planets and related bodies and hence I am prepared to write about the formation of the Solar System. However, in writing a book an author has first to consider the readership for whom it is intended and that is a problem with which I wrestled for some time. A complete deep scientific treatment of all aspects of the various theories would be unintelligible to most non-specialist readers and would just be a reproduction of what is already available in the scientific literature. An alternative approach, in which only verbal descriptions were given throughout, would be more readable but would lack credibility — many theories sound plausible enough when described in hand-waving fashion but wilt under close scientific scrutiny. So, to maximize the readership while maintaining scientific integrity, I have decided on a middle course. Fortunately the level of science needed to deal with most aspects of cosmogony can be understood by anyone with a fairly basic scientific background. The approach that has been adopted is to introduce equations here and there to provide scientific substance together with narrative to explain what they mean. In addition, for those who wish to delve more deeply into the subject, reference will be given to a small selection of books and papers in scientific journals — but I stress, these are *not* essential reading! Hopefully, this text alone will provide an account in a form that should both be understood by the non-expert reader and also be of interest to those with wider knowledge of astronomy or general science.

Introduction to the Second Edition

I do not believe in re-writing history so I have retained the Introduction to the First Edition in its original form — although I do not now stand behind everything I then wrote. As an example, not a very important one, the generally accepted age of the Universe is not the 15 billion years I quoted but closer to 13.5 billion years. Of more importance is that I have reneged on my promise not to write about the origin of the Universe and have done exactly that, although only as an introduction to the formation and evolution of the Solar System, especially of the Earth and all it contains, including life forms up to *homo sapiens* — us.[1] Such is the frailty of man!

However, the main justification for producing a second edition is not provided by my previous deficiencies but rather by developments in planetary science in the six years since the first edition was written. Pluto is no longer regarded as a regular planet but has been demoted to a new class of objects, of which it is not even the major member. Again, new observations of exoplanets, planets around other stars, have shown that they can be at large distances from their stars and also have large orbital inclinations, even to the extent of having retrograde orbits relative to the spin axes of the parent stars. These exoplanet observations have implications for the plausibility of some theories of planetary formation. Taken together with recent new hypotheses and simulations concerning the way in which the Solar System could have developed to its present form, this new material brings the second edition up-to-date at the time of publication although, in the nature of the subject, that condition may not last for long.

---

[1] *Time, Space, Stars and Man: The Story of the Big Bang* (Second Edition, 2013) Imperial College Press.

Prologue: The Dreamer

Gng lay on his back with his head cushioned on a bale of ferns. He was well away from the fire, on the windward side, so that the sky he saw was clear of the sparks that flew high into the air. His stomach was distended with aurochs meat, the product of the successful hunt that he and the other men of the tribe had carried out that day. The women too had done well, with a rich harvest of berries and roots that accompanied the meat in their gargantuan meal. These were the good days. He shivered with apprehension as he thought of the bad days that would soon come. Even with his bearskin cloak and leg covering he would feel the bitter cold. There would be many days when cold and hunger filled his mind to the exclusion of all else. These were times when the wild beasts were as desperate as the members of the tribe and last year two of the children had been taken.

He studied the pattern of lights in the sky that he had come to know so well. His imagination created pictures like those seen in a fire, but the pictures in the sky never changed. There is the snake, there the bear, the waterfall seemed particularly brilliant tonight and the hunter's spear is as clear as ever. Actually, there were some occasional changes. Many moons ago he had seen a bright light suddenly appear in the sky. The Moon had been eaten by the night so he could see that this light could make shadows. The brightness had lasted for two moons or so, gradually fading until it could no more be seen. Perhaps there was a great forest in the sky and a fire had raged in it and gradually died away. Such things were part of his experience. There were other changes that were less exciting but that were always happening. As each night passed so the sky lights twisted round, like a leaf on the end of a filament of a spider's web but always in one direction. They all twisted together so the patterns did not change. But, over the course of time, he had seen three lights that slowly moved amongst the others. What were the lights that kept their rigid patterns and what were those that travelled amongst them? Were those travellers like the old rogue males expelled from the herd by a new young dominant bull? No, that did not seem to fit.

He never spoke to the others about what he saw and what he thought. They knew that he was a little different — the name they had given him meant 'dream' or 'dreamer'. But he was a brave hunter and a respected member of the tribe so the difference was tolerated. Once, in a moment of rare tenderness, he had tried to explain to Nid, his woman, about what he saw in the sky. He could not find the words to express his thoughts and she comprehended nothing. She roughly pulled away from him, looked at him with a puzzled and troubled expression and then returned to suckling their latest infant. That was instinct — that she understood.

Gng did not know that he was a very important man. He had observed the night sky and tried to make sense of what he saw in terms of what he knew. He was the first astronomer.

Part I
General Background

Chapter 1

Theories Come and Theories Go

It's all kinds of old defunct theories, all sorts of old defunct beliefs, and things like that. It's not that they actually live on in us; they are simply lodged there and we cannot get rid of them.

Henrik Ibsen (1828–1906). *Ghosts* (1881)

1.1 What is Science?

The word 'science' is derived from the Latin *scientia* meaning 'knowledge'. In its modern usage science is a quest for knowledge and an understanding of the Universe and all that is within it. Knowledge grows with time; individual scientists learn from their predecessors and their work guides those that follow. Arguably the greatest scientist who has ever lived, Isaac Newton, recognized this debt to his predecessors by saying, "If I have seen further, it is by standing on the shoulders of giants."

All that Newton discovered is so much the accepted background of scientific endeavour today that, at least in astronomy and physics, what he did may now seem to be obvious and humdrum. Yet, in its day, it was spectacular. It was as though humankind, or at least those who could understand what Newton had done, had a veil moved from before their eyes so that all that was previously obscure was seen with a crystal-like clarity. The forces of nature that caused the Moon to go round the Earth and the Earth to go round the Sun were quantified. Forces that operated in the same way, but with different causes, could explain the way that electric charges attracted or repelled each other and also the behaviour of magnets. While all agree that Newton was a great man and his discovery of the law of gravity was a great discovery, can it be said that it was truth in some absolute sense? Apparently not, because three hundred years later

3

another famous scientist, Albert Einstein, showed that Newton's law of gravitational attraction was just an approximation and, to be *very* precise, one should use the Theory of General Relativity instead. It turns out that Newton's way of describing gravity is good enough for most purposes and the calculations that send spacecraft to distant solar-system bodies with hairline precision use Newton's equations rather than those of Einstein.

The example of gravitation is a good one for portraying one aspect of scientists' attitude to their work. Some of them are purely interested in theoretical matters, in that they just try to understand the way that nature works without necessarily having some practical motive to do so. One of the early deductions from the Theory of General Relativity is that light from a star, passing the edge of the Sun, will be deflected twice as much as would be suggested by Newton's gravitational theory. It was realised that if this could be shown to be true then General Relativity would get a tremendous boost in credibility. This prediction about the deflection of light was made by Einstein in 1915 while he was working in Berlin during the First World War. The observational confirmation that Einstein was right was made in 1919 by British teams of scientists, led by Sir Arthur Eddington and the Astronomer Royal, Sir Frank Dyson. They travelled to South America and West Africa to make observations during a solar eclipse, when starlight deflected by passage close to the Sun could be seen. The expedition was planned in 1918 while Britain and Germany were on opposite sides of a vicious and destructive war but their scientists could come together in their search for knowledge.

The demonstration that Einstein's prediction was right excited the scientific community, and even members of the general public who realised that something significant had happened even if they did not know quite what it was. Although important scientifically, this demonstration was *not* important in making a great impact on everyday life. However, the life that we live today *is* very much shaped by the science that has been done in the last 200 years. Experiments with 'Hertzian waves' in the latter half of the 19th century eventually led to radio, television, the mobile telephone and the internet.

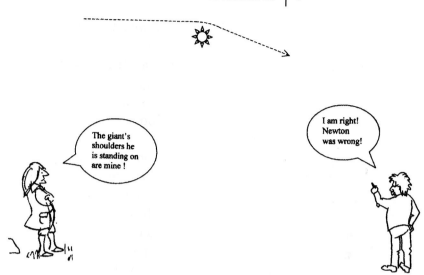

Figure 1.1   British teams of scientists observed that Einstein's predictions were right.

Einstein's Special Relativity Theory suggested the idea that matter can be turned into energy (and *vice-versa*) through the famous equation $E = mc^2$. The world has not been the same since the first large-scale demonstrations of the validity of that equation when atomic bombs were dropped on Hiroshima and Nagasaki in 1945. Curiosity about the way that electrons behave in semiconductor materials led to the electronics revolution that so dominates world economics. In fact, rather oddly, it sometimes seems that curiosity-driven research outperforms utility-driven research in terms of the usefulness of the outcome. When, in 1830, Michael Faraday waggled a magnet near a coil of wire and produced an electric current he was not conscious of the fact that he was pioneering a vast worldwide industry for generating electricity.

1.2 The Problem of Cosmogony

Although by no stretch of the imagination could one envisage any practical outcome from the deflection of a beam of light passing the Sun, at least it was possible to do the experiment to show that the prediction was true. There are other areas of science where there is limited opportunity for experimentation and one of those is the subject that we deal with here — technically referred to as *cosmogony*. The question we address is: How was the Solar System formed and how has it evolved since it formed? While there are various ideas in this field, everyone is agreed that the Solar System formed some 4,500 million years ago and that, since that time, it has undergone many changes of an irreversible kind. The concept of reversible and irreversible changes is quite fundamental in science. For a reversible change the system could, at least in principle, run backwards so that the past states of the system can be deduced from its present state. For example, we know where the Earth is in its orbit around the Sun and the laws of mechanics that govern its motion. We now imagine that the direction of motion of the Earth reversed so that it retraces its path or, in practice, we do the calculation corresponding to that reversal of motion. This enables us to find exactly where it was at times in the past — but not too far in the past, as even the Earth in its motion has undergone some irreversible processes. As an example of an irreversible process we imagine a large cubical evacuated chamber with a tap at each of its eight corners leading to a cylinder of gas. The tap is turned on at one of the corners and the chamber fills with gas. Once the gas has occupied the whole chamber there is no way of telling from which of the corners the gas came in. The event was irreversible and one cannot make the molecules of the gas reverse their motions and re-enter the cylinder from which corner they came. For an irreversible change the past state cannot be deduced from the present state.

If we cannot work backwards to find out how the Solar System began then what *can* we do? The answer is to try various models that are scientifically plausible to see whether or not they can give rise to a system like the Solar System today, or even one that might have evolved to give it. Taking this approach runs the risk that there would be a large number of models that could lead to the Solar System as we know it — but this turns out not to be the case. As we shall see, finding a model that gives anything like the Solar System has proved to be a very difficult exercise.

1.3 New Theories for Old

The history of science is littered with ideas that have held sway, that were eventually found to be flawed and were then replaced by some new idea. The lesson to be learnt from this is that no theory can ever be regarded as 'true'. There are two categories of theory — those that are plausible, and those that are implausible and therefore almost certainly wrong. Any theory in the first category is a candidate for the second whenever new observations or theoretical analysis throw doubt upon its conclusions. There is no shame in developing a theory that is eventually refuted. Rather, the generation and testing of new ideas must be regarded as an essential part of the process by which scientists gain the knowledge and understanding they seek. The Earth-centred theory of the structure of the Solar System due to Ptolemy, a 2nd Century Alexandrian Greek astronomer (see Section 3.5), was a useful model for the 1,400 years of its dominance and it agreed with what was known at the time. People's everyday experience suggested that the Earth was not moving because there was no sensation of movement such as one would have when walking or riding a horse. If the assumption that the Solar System was Earth-centred gave complicated motions of the planets with respect to the Earth then so be it — after all, there were no laws of motion known at that time that forbade such complication. When Copernicus introduced his heliocentric (Sun-centred) theory in the 16th century there were still no known laws of motion but the attraction of his idea was that, in terms of the planetary motions, it gave simplicity where the previous theory had given complication. It complied with a philosophical principle, known as *Occam's razor*, enunciated by the 14th century English Franciscan monk, William of Occam. The Latin expression of this principle — *Entia non sunt multiplicanda praeter necessitatem* — literally means, "Entities should not be multiplied beyond necessity", but is taken to mean, "If there are several theories that explain the facts then the simplest is to be preferred". It should be noted that, while simplicity is a virtue, the need to explain the facts is paramount.

A seeker after knowledge and understanding must be cautious about accepting ideas because they seem 'obvious' and fit in with everyday experience. That, after all, was the basis of Ptolemy's model. Other scientists of great stature have made similar errors. For example, Newton wrote in his great scientific treatise *Principia* (1687) as one of his 'Rules of Science':

"To the same natural effects, the same causes must be assigned."

As an example, he gave the light of the Sun and the light of a fire but we now know that these lights have very different causes. The heat of the Sun is produced by nuclear reactions while that of a fire comes from chemical changes produced by ignition due to rapid oxidation. Again, Einstein never accepted quantum mechanics, especially the uncertainty principle, a recipe for defining the fundamental limits of our possible knowledge. An implication of the uncertainty principle is that we can never precisely define the state of the Universe at any time and therefore we cannot predict what its future state will be. As Einstein wrote in a letter to Max Born, "I, at any rate, am convinced that *He* is not playing with dice." By *He* Einstein meant God.

The watchword in science is 'caution'. All claims must be examined critically in the light of current knowledge and one's mind should never be closed to the possibility that a theory could be wrong. The eminent Austrian, later British, philosopher Karl Popper (1902–1994) stated that, "Whenever a theory appears to you as the only possible one, take this as a sign that you have neither understood the theory nor the problem which it was intended to solve." Any acceptance must be that of the *pro tem plausibility* of an idea since the possibility of new knowledge and understanding to refute it must be kept in mind. We must beware of bandwagons and be prepared to use our own judgements; history tells us that bandwagons do not necessarily travel in the right direction!

Chapter 2

Measuring Atoms and the Universe

"...si parva licet comonere magnis..." [*"if one can compare small things with great"*]

Virgil (70–19 BC)

2.1 Measuring Things in Everyday Life

In life in general, and in science in particular, it is necessary to measure things. If we buy some cheese we expect to know how much it weighs and to pay accordingly. If we set out on a journey we want to know how long it is so that we can plan the trip and arrive at the required time — time being another quantity that we measure. In the everyday world the units used for measurement are ones to which we can relate. Different societies have devised different measuring systems — for example, the Imperial System, with pounds and feet, devised in Britain and used in a slightly modified form in the USA. However, most societies are now converting to the metric system, one of the great legacies of Napoleonic France. The kilogram (kg) is a very convenient unit of mass. One hundredth of a kilogram is the mass of a one-page letter in its envelope and one hundred kilograms is the mass of a very amply built man. A metre (m) is a length we can readily envisage; one hundredth of a metre, i.e. one centimetre, is roughly the thickness of a finger and one hundred metres is the length of the shortest sprint race in the Olympics. For longer distances the kilometre (km), one thousand metres, is a better unit so that the distance from London to Edinburgh is 658 km. Time has a variety of units, although for the scientist the basic unit is the second (s), approximately the time of a heartbeat to put it in familiar terms. The unit of time, at least from when time was first defined, was the day, the time interval between successive noon times, when the Sun is at its zenith. This was divided into hours, minutes and seconds to give a way of defining the time of day with sufficient precision for most human activities and to express periods of time with convenient magnitudes for particular purposes. We would find it difficult to comprehend the time of a 100 m athletics race as 0.0001157 days rather than 10 s and the time taken to cross the Atlantic in a ship as 518,400 seconds rather than 6 days!

Another quantity that needs to be measured in everyday life is temperature. The prevailing temperature indicates what kind of clothing needs to be worn. The baby's bath should be at a comfortable temperature and body temperature can be an important diagnostic indicator of health. In the UK and USA the Fahrenheit scale is still frequently used and understood by the population at large. On this scale the freezing point of water is 32°F and its boiling point 212°F. The rather more logical Celsius (centigrade) scale, with the freezing and boiling points of water as 0°C and 100°C respectively, is in general use in most of the world and is supplanting the Fahrenheit scale. However, even the Celsius scale is not logical enough for a scientist. Temperature is a measure of the energy of motion of the atoms in a substance. As the temperature of a gas is increased, so the gas atoms move around faster. The *Absolute or Kelvin scale* of temperature measurement (the unit, the kelvin, indicated by K) is defined such that the temperature is proportional to the mean energy of their motion. Atoms in a solid are fixed in a rigid arrangement but they do oscillate around some average position and, once again, the temperature is proportional to the mean energy of this vibrational motion. If there is no energy of motion then the temperature is zero on the Kelvin scale.[1] The increments corresponding to one degree are made the same as those of the Celsius scale. This gives 0 K = −273.2°C, the freezing point of water, 0°C = 273.2 K and the boiling point of water, 100°C = 373.2 K.

---

[1] To be precise, even at 0 K there is some residual energy, which physicists call *zero-point energy*. However, we can disregard this in our discussion.

2.2 Science and Everyday Life

Simple science that was developed until about 150 years ago was mostly about phenomena that played a part in everyday life. The laws of mechanics are ones that can be understood intuitively because they form part in our experience. Children playing 'catch' with a ball know nothing about the mathematics of parabolic motion but they understand instinctively how it works in practice. A footballer does not have to be a scientist to curl a ball into the corner of the net and the magical performances of great snooker players are based on experience, not scientific qualifications. Few people know about the intrinsic nature of light but everyone knows that you cannot see around corners. Understanding the way that the material world behaves has 'survival value' and so such understanding governs our instinctive behaviour.

A problem in understanding modern science is that it encompasses phenomena that are well outside the experiences of everyday life. We are not conscious of the presence of individual atoms, the size of the Universe does not impinge on our daily lives and the mechanical objects that form our environment do not move at a large fraction of the speed of light. Our usual scales for defining mass, distance and speed are inappropriate for dealing with the entities involved in present-day science and other scales are needed.

2.3 Small Things Beyond Our Ken

The idea of the atom originated with the 5th Century B.C. Greek, Leucippus. He imagined what would happen if one bisected a piece of matter over and over again making it smaller and smaller. He concluded that eventually an ultimate indivisible piece of matter would remain — and this was the atom. Leucippus assumed that every individual material could be reduced to the atomic state, so that we could have an atom of wood or an atom of stone, but now we know that these materials are themselves made of combinations of elemental atoms, such as oxygen and carbon. We also now know that the atom itself consists of even smaller components. An atom contains *protons* with a positive charge, *neutrons* with no charge but nearly the same mass as a proton and *electrons* with a negative charge equal in magnitude to that of the proton, but with a tiny fraction of the proton's mass. In some situations, when an atom breaks up, mysterious particles called *neutrinos* are produced which have no charge, possess energy and momentum and have an extremely tiny mass — once thought to be no mass at all. We cannot cope with the concept of a neutrino in a framework of everyday experience. There are even smaller particles, *quarks*, from which some atomic sub-particles are made but we shall not go further in that direction. What we have established is that there is the world of very tiny objects, a world that does not directly relate to everyday life.

If we write down the mass of a proton then, in decimal form, it is

$$0.00000000000000000000000000167 \, \text{kg},$$

a representation that is almost unintelligible. A slightly better representation is

$$\frac{1.67}{1,000,000,000,000,000,000,000,000,000} \, \text{kg}$$

but even this looks very clumsy and is very difficult to decipher. The divisor contains 27 zeros and it represents a product of twenty-seven 10s or, in other words, 10 to the power 27. We have a way of expressing this so that now we can write the mass of a proton as $1.67/10^{27}$ kg or, by a final transformation, 1.67×10^{-27} kg. This is now a far more succinct expression and with experience one can even begin to get a feel for what such a quantity means. In the same notation the mass of an electron is 9.11×10^{-31} kg so that about 1,835 electrons have the same mass as a proton.

Just as for mass, so atomic particles have extremely small sizes. A small atom has a diameter about 10^{-10} m, which means that a few million of them side by side will have the width of a dot over the letter 'i'. A proton is even smaller, 10^{-15} m in diameter, so that 100,000 of them will fit across an atom. Because the masses and linear dimensions of atoms and elementary particles are so small compared with the basic units, the kilogram and the metre, atomic and nuclear scientists have devised new units. Thus the *atomic mass unit* (amu) is roughly the mass of a proton but is defined as one-twelfth of the mass of a carbon atom.[2] For length, a convenient unit in nuclear physics is the *fermi* (fm), which is 10^{-15} m, about the size of a proton.

2.4 Measuring Things in the Solar System

Tiny atomic particles are important to scientists and, indeed, to technologists as well since much of the modern communications industry depends on the way that electrons behave. However, they have no obvious part to play in explaining how the Solar System and other planetary systems arose, nor are the small dimensions required to describe them relevant in this respect. So now we will move on to the world of the very large — at least by human standards.

The orbits of planets around the Sun are not circles but ellipses. Such an orbit is an oval shape, as shown in Figure 2.1, with the Sun displaced from the centre.

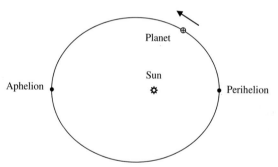

Figure 2.1 An elliptical planetary orbit.

Most planetary orbits are very close to circles, so close that if the orbit of the Earth had been shown in Figure 2.1 then the departure from a circle could barely have been detected visually. It is clear that as the planet goes round the Sun so its distance from the Sun is constantly changing. The point marked *perihelion* is when it is closest to the Sun and the point marked *aphelion* (pronounced afelion) is when

---

[2] Actually it is one-twelfth the mass of the most common isotope of carbon, carbon-12 (see Figure. 40.1).

it is furthest away. The terms perihelion and aphelion can also be used for the actual distances from the Sun.

The extent to which an ellipse departs from being a circle is measured by a quantity called its *eccentricity* denoted by e. One way of describing eccentricity is

$$e = \frac{\text{aphelion} - \text{perihelion}}{\text{aphelion} + \text{perihelion}} \qquad (2.1)$$

For a circular orbit e will be zero since the planet is always at the same distance from the Sun and the aphelion and perihelion will be equal. For an ellipse, e can be zero up to anything less than 1. For the Earth the aphelion is 1.521×10^8 km and the perihelion is 1.471×10^8 km which, put into (2.1), gives $e = 0.017$. Shown in Figure 2.2 is a selection of ellipses with a range of eccentricities together with the position of the Sun if the ellipses represented planetary orbits.

The eccentricity of an orbit describes its shape but not its size. The length of the line joining the aphelion to perihelion is called the *major axis* of the ellipse and half that distance, the *semi-major axis*,

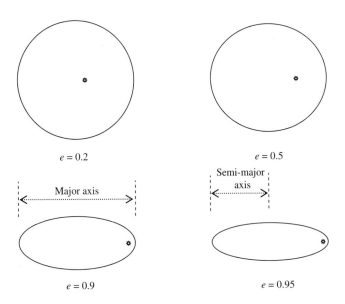

Figure 2.2 Elliptical orbits with various eccentricities. Also shown are the major axis and the semi-major axis.

is normally used to indicate the size of the ellipse. It is approximately the average distance between the planet and the Sun, with the average taken over a complete orbit.

An important distance in the Solar System is the semi-major axis of the Earth's orbit, which is 1.496×10^8 km. This distance is called the *astronomical unit* (au) and is a very useful unit for measurements in the Solar System so that, for example, the semi-major axis of Jupiter's orbit is 5.204 au.

The characteristics of the Earth and of its orbit are also used to define mass and time in many solar-system contexts. The Earth mass is 5.974×10^{24} kg; the most massive planet, Jupiter, has mass 317.8 Earth units and Mercury, the innermost and least massive planet, has mass 0.0553 Earth units. Similarly the time taken for the Earth to go round the Sun, the year, is a convenient unit for measuring the orbital periods of other planets — 11.86 years for Jupiter and 0.241 years for Mercury.

2.5 Large Things Beyond Our Ken

We have dealt with the kinds of measurements we need for discussing aspects of the Solar System. But, in terms of the Universe as a whole the Solar System is a tiny and insignificant entity. To measure distances in the Universe the metre, or even the astronomical unit, is far too small to be convenient. Instead we use the *light-year* (ly), the distance travelled by light in a year, as a convenient unit of measurement. Light travels at 2.998×10^8 m s^{-1} (metres per second) and there are 3.156×10^7 seconds in a year so

$$1 \text{ ly} = 3.998 \times 10^8 \times 3.156 \times 10^7 \text{km} = 9.46 \times 10^{15} \text{m} = 9.46 \times 10^{12} \text{km}.$$

The Sun is a member of the Milky Way galaxy, an island universe containing 10^{11} stars. Figure 2.3, a picture of a galaxy with the unromantic name NGC 6744, which is very similar to the Milky Way, shows what it is like. The distance across it is about 100,000 ly. In the Universe that we can detect with the most powerful telescopes, there are about 10^{11} galaxies. The nearest large one, the Andromeda galaxy, is close at hand, just 3 million ly away, while the furthest objects we

Figure 2.3 The galaxy NGC 6744 — very much like our own Milky Way galaxy.

can see are distant some 10^{10} ly. The Universe is expanding and these very distant objects are rushing away from us at a large fraction of the speed of light, our chosen scale for measuring speed in this context. A study of the structure of the Universe is fascinating — but is not part of our story.

Part II
Enlightenment

Chapter 3

Greek Offerings

Timeo Danaos — et dona ferentis [*"I fear the Greeks, even when they bring gifts"*]

Virgil (70–19 BC)

3.1 Even Before the Ancient Greeks

Three to four thousand years ago, while in most parts of Europe people lived in tribal communities or minor fiefdoms, great civilizations had grown up in Babylonia, Egypt, India and China. The very essence of a civilized society is that it has strong central control and that the individuals within it have varied roles that contribute to the whole structure of society. Astronomers, who studied the heavens for very practical reasons, played an important role in these early civilizations. The times for the sowing and harvesting of crops had to be determined and religious ceremonies had to take place at particular times of the year. Such needs dictated the establishment of a calendar, which, in its turn, required the study of the motions of the Moon and the stars. Another strong motivation for studying the heavens was to assist in astrological predictions. It was strongly believed in many early societies (and by some individuals even today!) that the fate of humankind is bound up in the motions of heavenly bodies. Eclipses and comets were the harbingers of great events and so it was important to know when they were coming and what they signified.

What the astronomers saw was great regularity and predictability in the heavens. They did not really comprehend the true nature of what they observed but they built up empirical rules that gave them powers of prediction — for example, of solar eclipses. There was just one phenomenon that departed from the smooth regular behaviour of all the other bodies they saw. These were the wandering stars, five of them, of which three made occasional looping motions in the sky while the other two could only be seen close to the Sun when it was below the horizon, just after sunset or before dawn.

These early astronomers were human and therefore they were curious — but we know little of any speculations they may have made about the nature of the objects they were observing. For that, we must move forward to ancient Greece, whose offerings we need not fear — just admire.

3.2 Plato and Aristotle

The Greeks were the first philosophers, in the archaic sense of that word which is synonymous with scientist today. Even in the twenty-first century the Physics department of the University of Edinburgh is named the Department of Natural Philosophy. However, the Greeks indulged in a curious mixture of philosophical concepts, in the modern sense, and of science, as we understand it. Thus, they had a concept of beauty and perfection and they felt that natural objects should conform to these concepts. Plato (c. 428–348 BC), who founded a great philosophical school, the Academy, thought that heavenly bodies had to be spherical because spheres were the perfect shape. For the same reason of perfection, he thought that all the motions of heavenly bodies had to be circles. Well, whatever the reasoning, he got the right answer for shapes, at least for all but very small astronomical bodies of which he would have been unaware in any case.

Greek philosophy did not confine itself to astronomical matters but here we shall just highlight those aspects of their thinking that advanced understanding about astronomy, in particular relating to the Solar System. A pupil of Plato, Aristotle (384–322 BC), had ideas ranging over the complete ambit of human thought. He conceived of a Universe structured in three parts. The outer bound of the Universe was a huge spherical shell in which the stars resided. Closer in was a region occupied by the Sun and the planets. Finally, right at the centre, were the Earth and the Moon. The placement of the Earth at the centre of the Universe, with everything else moving around it, is the most intuitive of beliefs since we have only to study the heavens to see that everything moves round the Earth — or so it seems.

3.3 Aristarchus — A Man Ahead of his Time

Aristarchus of Samos (310–230 BC) was an Alexandrian Greek with a rather untypical attitude to science in that he was prepared to do experiments. The Greeks had the practice of believing that if they came to a conclusion by the power of thought then that conclusion had to be true. For example, they had reasoned that heavy objects fall faster than light objects — which is certainly not true but they did not think it necessary to test the hypothesis.

Aristarchus was interested in knowing how far the Moon and Sun were from the Earth and he designed an experiment for finding out. He understood that the Moon was illuminated by the Sun and from geometrical considerations this meant that when half of the Moon was seen illuminated then the angle Earth–Moon–Sun had to be a right angle (Figure 3.1).

Aristarchus tried to measure the angle θ shown in the figure — presumably when the Moon could be seen in daylight on a clear cloudless day. This angle would not give the actual distances of the Sun and the Moon from Earth but it would give the ratio of the distances of the two bodies. What Aristarchus did not know was that

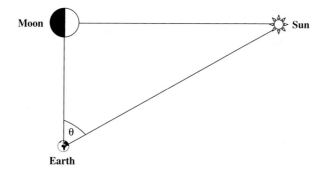

Figure 3.1   The Sun–Earth–Moon arrangement at half-Moon.

the distance of the Sun from Earth is much greater then the distance of the Moon so that the angle θ is only a little less than 90°. This means that, since small differences from 90° make a large difference to the estimated ratio, θ has to be measured very accurately. Aristarchus estimated the angle as approximately 87° and gave the ratio of the distances as 19 ± 1, much less than the true value of 390. The idea for the measurement was sound but the technology of the time was inadequate.

In addition to finding the relative distances of the Moon and the Sun, Aristarchus also made estimates of the relative sizes of the Earth, Moon and Sun. Since at the time of a total eclipse of the Sun the Moon just covers the Sun's disk it followed that the sizes of the two bodies were just proportional to their distances. Hence, according to Aristarchus, the Sun's radius was just 19 ± 1 times that of the Moon. To find the size of the Earth relative to the Moon Aristarchus used the phenomenon of a lunar eclipse when the Earth's shadow passes over the Moon's surface. He noted that the arc of the Earth's shadow was of considerably larger radius than that of the Moon itself and he estimated the ratio of the radii as about three. Actually, the radius of the Earth is nearly four times that of the Moon, but the idea was established that the Earth was considerably bigger than the Moon, and the Sun much bigger than both the other two bodies.

Aristarchus had another idea that was well ahead of its time. He proposed that the Sun and the stars were fixed in position and that the Earth moved in a circular path around the Sun. The idea that the Earth moved round the Sun would not be raised again for another 1,800 years and the general acceptance of that proposition would take even longer.

3.4 Eratosthenes — The Man who Measured the Earth

We have seen that the Greeks understood that the Earth is spherical in shape so, naturally they would wonder how big it was. The answer to that question was provided by another Alexandrian Greek, Eratosthenes (276–195 BC). It was well known by people local to that area, that at mid-day on the day of the summer solstice,

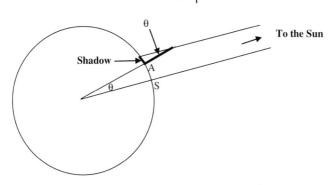

Figure 3.2 Light from the Sun shining straight down a well at Syene (S) but casting a shadow from a tower in Alexandria (A).

when the Sun was at its furthest northern position, it shone straight down a well at Syene (near modern Aswan). This meant that the Sun was at the zenith, i.e. directly overhead. Alexandria is due north of Syene so the Sun was not directly overhead but shone down at an angle to the vertical.

In Figure 3.2 the Sun's rays point straight down the well at S, the location of Syene. In Alexandria, located at A, a tall tower throws a shadow the length of which, relative to the height of the tower gives the angle θ. This angle is the difference in latitude between Alexandria and Syene. Eratosthenes estimated this angle to be 1/50th of a complete rotation so that the circumference of the Earth was just 50 times the distance between Syene and Alexandria. Long distances in those days were measured by paces; there were people trained to make a standard pace just as soldiers have always done. A Roman soldier made 1,000 paces to a Roman mile (a word based on the Latin *millia* meaning 'thousands'), where a pace corresponded to two steps in the Roman interpretation. In British infantry regiments a standard pace (single step) is 30 inches, except for light infantry regiments where it is 27 inches; if Roman soldiers had used a 30 inch step then the Roman mile would have been some 5% less than a British mile. The professional pacer found the distance from Syene to Alexandria to be some 5,000 stadia so Eratosthenes estimated the circumference of the Earth to be 250,000 stadia. The stadium is believed to have been about 180 m, giving the radius of the Earth as

7,200 km according to Eratosthenes — about 13% larger than the actual value of just under 6,400 km.

3.5 Ptolemy and the Geocentric Solar System

We all sense the world from the viewpoint of our own position in it, which is a statement of the obvious. Here you see a tree, there a house and above both of them a bird is circling in the sky. Since you have human imagination you can deduce what the scene would appear to be from the viewpoint of a distant hill or perhaps of the bird — but it is an effort. What is true for one person is true for humankind as a whole. We see the Universe from a stable unmoving platform, the Earth, and all other bodies, the Sun, the Moon, the planets and the stars move around it. For the ancient Greeks to have thought that the Earth was moving would have been almost impossible. Motion was something that was felt by the mover, walking or being transported by horse, wagon or ship. Clearly the Earth did not move.

Figure 3.3.   Ptolemy.

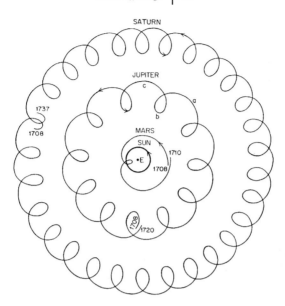

Figure 3.4   Motions of Mars, Jupiter and Saturn as seen from Earth.

This geocentric, i.e. Earth-centred, view of the Universe was supported by the motions of the Moon, Sun and stars that all moved in smooth circular motions in the sky. These motions agreed with a proposition put forward by Aristotle that the motions of all heavenly bodies were in circles at a constant speed. The exceptions to this harmonious picture were the planets that generally moved in an easterly direction but also, periodically, made great looping motions in the sky (Figure 3.4). Indeed this is why they had been given the name planets, derived from the Greek word meaning 'wanderers'.

It was yet another Alexandrian Greek, the astronomer Ptolemy (100–170 AD) who brought order to the motions of the planets. Since simple circular motion at a constant speed could not provide an explanation for their motion he devised a scheme involving two circular motions at a constant speed — illustrated in Figure 3.5. The first motion in a circle at constant speed was of a point called the *deferent*. The planet then moved round the deferent in a circle at constant speed, this path being called an *epicycle*. This model reproduced the motions of the planets quite well, certainly to within the

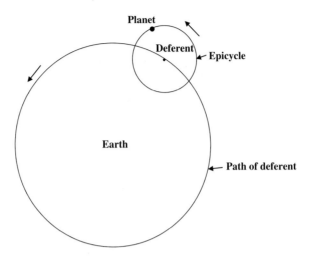

Figure 3.5   Ptolemy's description of planetary motion.

accuracy that the planetary motions were known at that time. There were also some rules that the motions had to obey. Since Venus and Mercury are always seen close to the Sun the deferents of these planets had always to be on the Earth–Sun line. Again, to explain the motions of the planets outside the Earth's orbit, the superior planets, the line joining the deferent to the planet had always to be parallel to the Earth–Sun line. These were complicated rules but it has to be remembered that there were no laws of mechanics known at that time and Ptolemy's model at least had the virtue that it was a systematic description of planetary motions. Indeed this was the model that was to be generally accepted without question for the next 1,400 years.

Chapter 4

The Shoulders of Giants

We are dwarfs on the shoulders of giants, so that we can see more than they, and things at a greater distance, not by virtue of any sharpness of sight on our part, or any physical distinction, but because we are carried up and raised high by their giant size.

Bernard of Chartres (c. 1130)

4.1 The Refugees

The beginning of Roman Empire can be traced back to the republic established in 510 BC. For several hundred years the Romans occupied large parts of Europe, North Africa and the Middle-East. In general, especially by comparison with their Greek contemporaries, they were not noted for their scientific or philosophical contributions but they were skilful administrators and they established a legal code, Roman Law, a novel institution in many of their conquered territories. They were extremely competent civil engineers, especially skilled in the construction of magnificent buildings, aqueducts, bridges and roads, and the routes of the roads they built are still followed by many modern highways. They were also adept in military technology and fully exploiting the resources, both in material and manpower, of the regions they conquered. Thus legions recruited in Syria manned Hadrian's Wall along the northern border of their British territories and legions recruited in Britain served in other parts of the Empire.

The Empire was often beset with turmoil and intrigue as various factions and individuals sought to become dominant. Eventually, in 395 AD the Empire split into two parts, the Western Roman Empire centred on Ravenna (since Rome came under frequent barbarian attack) and the Eastern Roman Empire with capital Constantinople.

Both Empires had Christianity as the state religion; Constantine, who died in York in 337 AD was the first Christian emperor. The last emperor of the Western Roman Empire was overthrown by the German chieftain Odoacer in 476 AD and thereafter the heritage of Rome was confined to the Eastern Empire.

After the birth of Islam in the seventh century AD the Eastern Empire was greatly influenced by the poetry, science and philosophy of the neighbouring Arab nations. However, after the Turks occupied Constantinople in 1453, many Christian scholars fled eastwards back to an Italy that, by this time, existed as collection of individual kingdoms within all of which there was a strong influence of the Roman Catholic Church. The renaissance, a scientific, artistic and literary revival, had begun in Italy in the 14th century but this received a great fillip from the arrival of cultured and educated refugees from the east. This is the background for a new surge in astronomical understanding, especially about the Solar System.

4.2 Nicolaus Copernicus and a Heliocentric Solar System

Nicolaus Copernicus (1473–1543) was a Polish cleric who spent some time as a professor of mathematics and astronomy in Bologna and Rome and also practiced medicine. In the 15th century, and indeed for some time thereafter, it was not uncommon for learned men to extend their studies over many apparently disparate fields, very unlike the situation today. The vast explosion of knowledge in the last two centuries has, perforce, led to increasing specialization so that, for example, the average physicist would not be very familiar with all details of developments in physics outside his or her immediate field of interest.

For most of his life Copernicus served the church in Warmia in Poland, about 150 kilometres from his native city, Torun. He became interested in the motions of the planets and he had available translations of Ptolemy's *Almagest*, the 13 books containing the accumulated astronomical knowledge of 2nd century AD. Copernicus constructed tables of planetary motions and, from the better available observational data, he concluded that Ptolemy's description of geocentric planetary

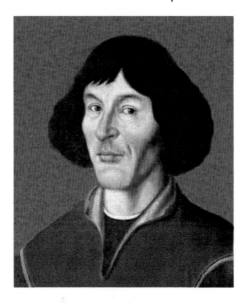

Figure 4.1 Nicolaus Copernicus.

motion was unsatisfactory apart from being rather complicated, which is a negative feature of any theory, then as now.

Using the improved observations, he developed a heliocentric model in which the planets all circled the Sun. However, like Ptolemy before him, he was reluctant to abandon the concept that planetary motions had to be based on circular motion at a constant speed. The observations required that the angular speed of planets in their motions around the Sun had to vary slightly and he satisfied this requirement by having the Sun displaced from the centre of the circular motion. When the planet was closer to the Sun then it would have a greater angular speed, which is just what was needed. Even with this modification there were still discrepancies between the motions of his model planets and the observations, so he introduced small epicycles to improve the agreement, although the required epicycles were tiny compared with those proposed by Ptolemy.

The church was aware of his astronomical work and was either supportive or, more likely, was perhaps just disinterested in what he was doing. Copernicus delayed the publication of his work for about 20 years and it eventually appeared in the treatise *De Revolutionibus*

Orbium Coelestium (*On the Revolution of the Heavenly Spheres*), which he dedicated to Pope Paul III. Just after the book was published Copernicus died — it was said that he received the printed copy on his deathbed. Anyway, he did not see the storm that his work was destined to stir up by the end of the century.

4.3 Tycho Brahe — The Man with a Golden Nose

Tycho Brahe (1546–1601) was from a noble family with close connections to the Danish royal family. While at the University of Rostock he quarrelled with a fellow student and in the ensuing duel part of his nose was sliced off — after which he wore a false nose made of gold and silver. He was a quick-tempered and arrogant man and these attributes were greatly to influence his subsequent career.

Tycho was interested in astronomy from an early age. In 1572 a very bright supernova appeared and he observed it over an eighteen-month period. The king, Frederick II, was very impressed by the

Figure 4.2.   Tycho Brahe.

young Tycho and offered to support his work on a grand scale. He made available to him the island of Hven, between Denmark and Sweden, and the resources to construct and run a large observatory. This observatory, called Uraniborg, was equipped with line-of-sight instruments based on very large quadrants for measuring angles, designed by Tycho himself. The engraving in Figure 4.3 shows one of these quadrants mounted against a wall of the observatory. The picture painted on the wall depicts Tycho, but the man himself can be seen at the right-hand margin of the engraving taking an observation. One assistant is recording the time from a clock while another assistant, seated at the table, is writing down the observations.

Figure 4.3   Tycho Brahe's quadrant.

When Tycho was making his observations the heliocentric model of the Solar System, as proposed by Copernicus, was well known and it might be thought that its clear superiority over the Ptolemaic system would have recommended it to all practicing astronomers. This was not so. The general acceptance of the geocentric model was so well established that there was considerable resistance to change. Lest we be critical of this conservatism of a bygone age we should note that, even today, there is a reluctance to abandon old ideas in favour of new ones — no matter how strong the evidence. Tycho himself adopted a hybrid model in which all the planets other than the Earth orbited the Sun, but with the Sun and the Moon orbiting the Earth. Actually, this correctly gave all the relative motions of the bodies, which was all one needed in the absence of any scientific theory to explain how they all moved. For that, one had to wait more than one hundred years.

Tycho ran into trouble when his patron, King Frederick II, died in 1588. The residents of the island of Hven were Tycho's tenants and he treated them very badly, a consequence of his arrogance and disdain for those he considered his inferiors. Frederick's successor, his son Christian, was less tolerant of Tycho's bad behaviour than his father had been and, in addition, he felt that Tycho took for granted the generous support he had been given and for which he showed very little appreciation. Eventually Tycho had to leave Hven and in 1599 he joined the court of Rudolph II in Prague as the Imperial Mathematician. This was a fortunate move. While in Prague he compiled tables of planetary motions, based on his years of accurate observations, assisted by a very able young man, Johannes Kepler. It was this young man who was to take the next significant step in increasing understanding of the nature of the Solar System.

4.4 Johannes Kepler — A Scientific and Mathematical Genius

Johannes Kepler (1571–1630) came from a poor family but, due to an enlightened government in Württemberg where he was born, he was able to receive an education. He first became interested in the

Figure 4.4 Johannes Kepler.

Copernican model when he was a student at Tübingen and some time later, when he taught mathematics at Graz, he wrote a book, *Mysterium Cosmographicum* (*Cosmic Mystery*), in support of the heliocentric theory. While he supported the general idea that all planets, including the Earth, orbited the Sun he was not happy with the description given by Copernicus that required small epicycles. He felt instinctively that the orbits should be a simple curve of some kind; unlike Copernicus he was not obsessed by the idea, originating with Plato, that all orbits should be based on circular motions.

Kepler sent a copy of his book to Tycho Brahe and this began a correspondence that eventually, in 1599, led to Tycho inviting Kepler to be his assistant at a new observatory that was being constructed at Benatek. Kepler readily accepted the offer — the thought of being able to analyse the superb data collected by Tycho Brahe was more than he could resist. In the event he was disappointed as Tycho gave him limited access to his data, but when Tycho died the data became his and he was able to begin his great project of analysing exactly how planets moved. The key to this work was the motion of Mars, the well-observed planet that most departed from circular motion.

After eight years of effort Kepler discovered the first two of his three laws of planetary motion; the third law took another nine years to be formulated. These laws are:

1. Planets move in elliptical orbits with the Sun at one focus.
2. The radius vector sweeps out equal areas in equal time.
3. The square of the period is proportional to the cube of the mean distance.

The first two laws, together with the determination of the shape of the orbit of Mars, were published by Kepler in *Astronomica Nova* (*New Astronomy*) in 1609 and the third law was given in his book *Harmonices Mundi* (*The Harmony of the World*) in 1619.

In retrospect it seems rather surprising that Kepler took so long in finding that the orbits were ellipses. He was a skilled geometrician and the ellipse was a well-known geometrical shape.

Although Kepler is best known for his laws of planetary motion he made contributions in many other areas of science. He was the first to propose that tides were due to the action of the Moon — although his great contemporary, Galileo, disagreed. He also knew that the Sun spun about its axis, a motion that he described in *Astronomica Nova*. He was interested in understanding how a telescope worked so he investigated the way that images were formed, including image formation by the eye. He also gave practical rules for designing eyeglasses for long and short sight. His optical work was published in two books, *Dioptrice* (the first word of a very long title) about refraction and other aspects of optics and *Astronomiae Pars Optica* (*The Optical Aspects of Astronomy*). He also understood how two eyes, giving slightly different views, could give depth to perception and suggested how the distances of stars could be found by observing them from points separated by a diameter in the Earth's orbit. The attempts he made to measure the distances of stars were unsuccessful because the instruments he used were not good enough, but the principle is sound and has been used in modern times with great success.

4.5 Galileo Galilei — Observation versus Faith

Galileo Galilei (1562–1642), a contemporary of Kepler, was appointed as professor of mathematics in Pisa at the age of 25 and also taught at Padua university for 18 years. He was interested in mechanics and also in planetary motion. Kepler sent him a copy of his *Epitome of the Copernican Astronomy* and in response Galileo indicated that he too favoured the heliocentric model.

In 1600 an event occurred that was to have important repercussions for Galileo. An Italian Renaissance philosopher and Dominican monk, Giordano Bruno, put forward the proposition that other stars were like the Sun and hence would have accompanying planets occupied by other races of men. This threw down a distinct challenge to the Church and its doctrine of the central role of mankind, moulded in God's image and with a unique relationship to the deity. Bruno was also heretical in other respects — for example, he did not accept the divinity of Jesus. Bruno's proposition, that mankind on Earth was not a unique example of its kind, was the last straw; Bruno was brought before the Inquisition but refused to retract his views and

Figure 4.5 Galileo Galilei.

paid for his steadfastness, or obstinacy, with his life. The source of the trouble, as far as the Church was concerned, was the Copernican model that demoted man's home, the Earth, to just one planet of many, and its view of *De Revolutionibus* underwent a complete change. In place of support, or at least tolerance, the work was added to the *Index Librorum Prohibitorum*, the list of books that Catholics were forbidden to read, and it was joined on the list by Kepler's writings. In case it should be thought that the Catholic Church was alone in being against the heliocentric theory it should be stressed that it was also anathema to the Lutheran Church that was dominant in the north of Europe. However, the Lutheran Church did not have such power and influence in northern states as the Catholic Church wielded in the south, which meant that Kepler could pursue his ideas unmolested. This was not so for Galileo.

In 1608, following its invention in Flanders by a lens-maker, Hans Lippershey, Galileo made a telescope. He was encouraged to do this by the Venetian Senate. With even a low-power telescope, ships could be seen approaching the harbour while still two hours sailing time away, which had both commercial and military value. However, Galileo saw the telescope as a powerful tool for astronomy and he used it to great effect. He saw lunar mountains and made estimates of their heights by the length of the shadows they cast. He also discovered four large satellites orbiting the planet Jupiter — Io, Europa, Ganymede and Callisto — now known collectively in his honour as the *Galilean satellites*. Seeing these satellites around Jupiter, like a miniature version of the Solar System, reinforced Galileo's belief in the essential correctness of the Copernican model. Another important observation, in 1610, was of the planet Saturn that he saw with a pair of appendages looking like ears. We now know that what he was looking at were Saturn's rings but the quality of his telescope was too poor to show a clear image. Galileo was astonished when in 1612 the appendages disappeared; the rings were edge-on to his line of sight but he did not know that.

None of these observations had relevance to the controversy concerning the geocentric and heliocentric alternative models of the

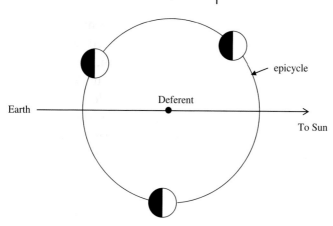

Figure 4.6   Venus as seen from earth according to Ptolemy's model.

Solar System. But there was one observation made by Galileo that had direct relevance and that was of the changing phases of Venus. According to Ptolemy's model, with the deferent of Venus's orbit always on the Earth–Sun line as seen in Figure 4.6, it would always be the rear of Venus that was illuminated by the Sun so that only a part of the edge of Venus would appear to be lit up and could thus only appear as a crescent.

The Copernican model gives a different outcome, shown in Figure 4.7. Both Venus and the Earth are now orbiting the Sun, with Venus in the inner orbit. Sometimes Venus is between the Earth and the Sun and Venus is seen in a crescent phase. When this happens Venus is close to the Earth so the crescent is large. At other times Venus and the Earth are on the opposite sides of the Sun. In this case a whole circular face of Venus is illuminated and a 'full' Venus is seen. When that occurs Venus is at its greatest distance from the Earth and Venus appears small.

Galileo's observations showed clearly that the Copernican theory was the correct one but he was a devout man (despite siring several illegitimate children) and the Church had set its face against the heliocentric model — so what could he do?

Galileo decided to try to convince the world at large of the correctness of the Copernican theory without unduly offending the

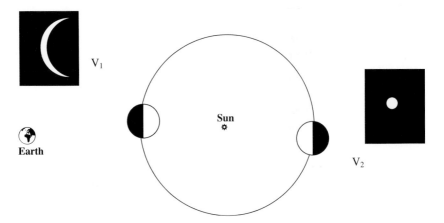

Figure 4.7   Venus as seen from earth according to the Copernican model.

Church. In 1632 he wrote a book, *Dialogue on Two World Systems,* in which two individuals, Simplicius and Salvatio, argued the respective merits of the geocentric and heliocentric models. They addressed their arguments to Sagredo, an intelligent layman who asked questions and so acted as a catalyst for the discussion. The advocate of the geocentric model, Simplicius, as his name almost suggests, argued very ineffectively compared with the lucid and compelling Salvatio. Galileo's device of trying to protect himself by appearing to present an impartial argument did not work. The Pope, Urban III, was furious and Galileo was ordered to appear before the Inquisition. He was no Bruno and sensibly gave way to brute force and humbly recanted and proclaimed that he sincerely believed that the Earth was stationary and the centre around which all other bodies moved. There is a story, probably apocryphal, that after his recantation he muttered under his breath, *Eppur si muove* — "but still it moves".

Because of his eminence Galileo was not subjected to ill-treatment nor was he imprisoned but he was under virtual house arrest for the rest of his life. His public humiliation was a sad affair but it was probably the last time that the new scientific method of deduction based on observation was challenged on an irrational basis.

4.6 Isaac Newton — And All was Light

On Christmas day in the year that Galileo died the greatest scientific genius of all time, Isaac Newton (1642–1727), was born. The poet Alexander Pope wrote,

> Nature, and Nature's laws lay hid in night:
> God said, *Let Newton be!* And all was light.

Newton formulated 'Newton's laws of Motion', he studied light and various aspects of hydrostatics and hydrodynamics and he developed the branch of mathematics which he called *fluxions* but is now called *calculus*. His greatest contribution to astronomy was the formulation of the inverse-square law of gravitation that states that, "the gravitational force between two point masses is proportion to the product of the masses divided by the square of the distance between them". One of his laws of mechanics specified that a body could only change its speed or direction if it had a force acting on it.

Figure 4.8   Isaac Newton.

Since the Moon in its orbit constantly changes direction it must have a constant force acting on it and the Earth was the obvious source of that force. By this kind of argument, and perhaps from watching apples fall off trees, he was led to his law of gravitational force. He was able to demonstrate that Kepler's three laws of planetary motion followed from the law of gravitational attraction, in particular that the orbits of planetary bodies round the Sun or satellites around planets were ellipses.

Newton described his theory of gravitation, and much else he had worked on, in a major publication, the *Principia*,[1] which was published in 1687. It took 15 years to write and he had to be persuaded to publish it by his good friend Edmund Halley, whose name is associated with a well-known comet. Halley even paid for the cost of the publication. It can truly be said that Newton's work represented the greatest ever watershed in science, in particular in understanding planetary dynamics. From Newton onwards the essential structure of the Solar System was clearly established and clearly understood.

Newton was an irascible and quarrelsome individual who made many scientific enemies. Robert Hooke, an eminent contemporary scientist with great achievements to his credit, claimed to have considered an inverse-square law for gravitational attraction before Newton had done so and to this remark Newton reacted with characteristic vehemence. It is even said that when he became President of the Royal Society he arranged for the only existing portrait of Hooke to be destroyed so that now we have no idea, except from verbal descriptions, of what that great scientist looked like. More serious was Newton's quarrel with the German mathematician Leibnitz over who had first invented calculus. The truth of the matter is certainly that they had done so independently although, in fact, it is Leibnitz's mathematical notion that we use today.

Not all of Newton's scientific activities were fruitful. He devoted many years to alchemy — an attempt to change base metals into gold

---

[1]The full title is *Philosophiæ Naturalis Principia Mathematica* (*Mathematical Principles of Natural Philosophy*).

and to create an elixir of life. For all his genius, he was a child of his time and alchemy was still regarded as a marginally-legitimate branch of science — although looked down on by many of Newton's contemporaries. The last years of his life were spent as Master of the Royal Mint. It was characteristic of the man that he pursued counterfeiters relentlessly and many of them suffered death and imprisonment thereby. The greatest of all scientists, maybe, but not the greatest of all men!

Part III

The Solar System:
Features and Problems

Chapter 5

The Sun and the Planets

Our souls, whose faculties can comprehend
The wondrous Architecture of the world:
And measure every wand'ring planet's course
Still climbing after knowledge infinite, ...

Christopher Marlowe (1564–1593)

5.1 The Sun

In this and the following two chapters we shall be describing the Solar System following an approximate hierarchy of sizes of its constituent bodies, starting with its largest and dominant member, the Sun, a typical main-sequence star[1] that accounts for 99.86% of the total mass of the system. With a mass of 1.989×10^{30} kg and a radius of 6.955×10^5 km it has about 1,000 times of both the mass and volume of the largest planet, Jupiter.

The composition of the Sun by mass is about 71% hydrogen, 27% helium with 1–2% of heavier elements, which is typical of the contents of the Universe as a whole. It has two characteristics that we shall see are important when it comes to considering theories of the origin of the Solar System. The first is that it spins on its axis quite slowly, with a period of approximately 25 days, most quickly at the equator with the spin rate decreasing (period increasing) with latitude. The second important characteristic is that its spin axis is inclined at about 7° to the mean plane of the Solar System, as defined by the planetary orbits. This angle is large enough to be a problem for some theories and small enough to be a problem for others.

---

[1]A main-sequence star is one that generates energy by converting hydrogen to helium in its core (see Section 17.2).

5.2 Types of Planet and their Distribution

The next largest members of the Solar System are the planets, eight in all, and clearly divided into two groups, the large *major planets*, Jupiter, Saturn, Uranus and Neptune and the much smaller *terrestrial planets*, Mercury, Venus, Earth and Mars. The terrestrial planets are spheres of mainly silicate rock with iron cores and all except Mercury possess atmospheres, which are minor components. By contrast, the major planets are mainly gaseous but with silicate-iron cores. The masses, radii and mean densities of the planets are shown in Table 5.1.

The two groups of planets are also distinguished by their locations within the Solar System. The terrestrial planets are closely spaced in the inner part of the system while the major planets are much further out and well separated from each other (Figure 5.1). With the exceptions of Mars and Mercury, planetary orbits are close to circular. They are also close to being coplanar; the standard plane for orbits is the *ecliptic*, the plane of the Earth's orbit, and the plane of the orbit of any other planet is given by its *inclination* to the ecliptic. The characteristics of the planetary orbits are given in Table 5.2, from which it will be seen that Mercury, in addition to having the greatest eccentricity also has the greatest inclination.

With the advent of the space age our detailed knowledge of the planets has greatly increased. However, the description of the eight

Table 5.1   The physical characteristics of the planets (Earth mass = 5.974×10^{24} kg)

Planet	Mass (Earth units)	Diameter (kilometres)	Density (kilograms per cubic metre)
Mercury	0.0533	4,879	5,427
Venus	0.8150	12,104	5,243
Earth	1.0000	12,756	5,515
Mars	0.1074	6,794	3,933
Jupiter	317.8	142,984	1,326
Saturn	95.16	120,536	687
Uranus	14.5	51,118	1,270
Neptune	17.2	48,400	1,638

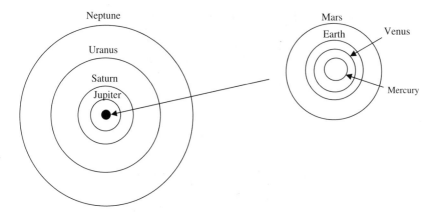

Figure 5.1 The planetary orbits. The orbits of the terrestrial planets are contained in the solid black circle.

Table 5.2 The orbital characteristics of the planets

Planet	Semi-major axis (au)	Eccentricity	Inclination (°)	Orbital period (years)
Mercury	0.387	0.2056	7.0	0.2409
Venus	0.723	0.0068	3.4	0.6152
Earth	1.000	0.017	0.0	1.0000
Mars	1.524	0.093	1.8	1.8809
Jupiter	5.203	0.048	1.3	11.8623
Saturn	9.539	0.056	2.5	29.458
Uranus	19.19	0.047	0.8	84.01
Neptune	30.07	0.0086	1.8	164.79

planets that will be given here will not be exhaustive, but sufficient for the purpose of considering the mechanisms by which they may have been formed.

5.3 The Major Planets

From Table 5.1 we can see that the major planets seem to fall into two separate pairs, distinguished by their sizes and related masses. The larger pair, Jupiter and Saturn, are the *gas giants*, with a gaseous composition by volume of hydrogen (~ 90%) and helium (~ 10%) with

Figure 5.2 Jupiter's Red Spot, a massive storm system, is seen in the lower-central part of the figure.

small amounts of other constituents, such as water methane and ammonia. By contrast the smaller pair, Uranus and Neptune, contain only about 40% of hydrogen-plus-helium and large amounts of volatile materials, such as water and methane that would be in the form of ices at low temperature (although not in the hot interiors of the planets) which gives rise to the name *ice giants* for this pair of planets.

We can take the prototype gas giant as Jupiter (Figure 5.2), the largest planet, with mass two-and-a-half times that of all other planetary masses combined and with 318 times the mass of the Earth. While mostly consisting of hydrogen and helium, it has minor components of molecular species made up of various combinations of hydrogen, carbon, nitrogen and sulphur and its silicate-iron core probably has a mass of between 5 and 10 Earth masses. Its banded surface contains oval swirls of material, one of which is the *Great Red Spot*, a large oval feature several times bigger than the Earth. This is a storm, like a hurricane on Earth, which has been seen on Jupiter for more than 350 years. Smaller ovals are also storms but with much shorter durations of a year or so. Jupiter has a spin period of 9 hours 55 minutes — the fastest spin rate of any large body in the Solar System. An interesting characteristic of all planets is the tilt of their spin axes, the angles made by the spin axes with the normal (perpendicular) to their orbital planes. This is small, just 3°, for

Table 5.3 Tilts of the planetary spin axes

Planet	Axial tilt (°)
Mercury	0.0
Venus	177.4[†]
Earth	23.4
Mars	25.2
Jupiter	3.1
Saturn	26.7
Uranus	97.8[†]
Neptune	28.3

[†]The spins of these planets are retrograde

Jupiter but, as can be seen in Table 5.3 is much larger for some of the other planets.

Saturn's surface features are similar to those of Jupiter, although with no feature as long-lasting as the Great Red Spot. Saturn's spin period varies from 10 hours 15 minutes at the equator to 10 hours 38 minutes closer to the poles. Its material is highly concentrated towards the centre, with outer material being quite diffuse, giving it a very low overall density, 687 kg m^{-3}, just under over that of Jupiter. The combination of its fast spin and density distribution leads to discernable flattening along its spin axis.

The ice giants are considerably smaller and less massive than Jupiter. The colour of Uranus, greenish blue, is due to the presence of methane, CH_4, in its atmosphere. Its spin period is 17 hours at the equator but *decreases* to 15 hours near the poles; this increase of spin rate with higher latitudes differs from the pattern in Jupiter and Saturn. The most curious feature of Uranus is the tilt of its spin axis, 98°, which means that its spin axis is only 8° from its orbital plane and that the spin, unlike the majority of rotations in the Solar System, is retrograde, i.e. clockwise as viewed from the north. This strange arrangement is almost certainly related to some event in its history.

Neptune is similar to Uranus, although its greater mass and smaller radius gives it a significantly greater density. The tilt of its spin

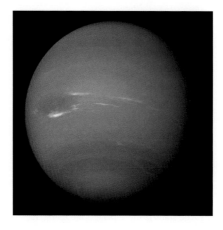

Figure 5.3 Neptune with the Great Dark Spot visible at centre-left.

axis, 28.3°, is slightly greater than that of the Earth, Mars and Saturn — large, but not unusually so. Its major surface feature is the *Great Dark Spot*, a great storm system, which is visible in Figure. 5.3.

5.4 The Terrestrial Planets

As for the major planets, the terrestrial planets seem naturally to fall into two distinct pairs — the larger and more massive Earth and Venus and the smaller pair Mars and Mercury.

The largest member, the Earth, has the primary characteristic that over a large part of its surface the temperature is such that water is in a liquid form. About 70 percent of the surface is covered by water, as liquid or ice, and atmospheric processes causes precipitation on the land regions so that some water is present even there. The presence of life owes much to the presence of liquid water, and advanced life forms to the evolution of an atmosphere which contains 21 percent oxygen, nearly all the remainder being nitrogen. It is worth noting that the fortunate temperature range that gives so much liquid water is due to the *greenhouse effect*, whereby some gases in the atmosphere — predominantly water vapour and carbon dioxide — are transparent to the shorter wavelengths of light emitted by the Sun, which cause the Earth to heat up, while being almost

Figure 5.4 A radar view of the surface of Venus

opaque to the longer wavelengths emitted by the much cooler Earth's surface. Without this effect the average Earth temperature would be 33° lower than it is now and most water would be in the form of ice. It would have been possible, but much more difficult, for life to have evolved in that environment. The Earth's orbit is almost circular and the 23½° tilt of its axis gives the seasonable effects with which we are familiar.

Venus is unlike the Earth in most characteristics. The thick carbon dioxide atmosphere gives a surface pressure 100 times that on Earth and it is opaque to visible light. Figure 5.4 shows a radar image taken by an orbiting spacecraft. There are three elevated regions that, if there were seas on Venus, would be continents but, in fact, Venus is very arid. Because of its proximity to the Sun the early atmosphere would have contained a great deal of water vapour. Ultraviolet light from the Sun would have broken up water, H_2O, into OH+H and the light gas hydrogen then escaped from the atmosphere. The hydroxyl radicals, OH, combined in pairs to give H_2O + O and the oxygen would have oxidized some local material. In this way most of the water would gradually be lost; the water component of Venus is about a one-thousandth of that on Earth.

The surface temperature of Venus, 730 K, is similar to the maximum on Mercury, which is much closer to the Sun. Because the atmosphere is dominantly carbon dioxide the greenhouse effect acts very powerfully and the surface temperature has to be very high before transmission of energy outwards balances that inwards, coming from the Sun.

The next planet in order of size, Mars, named after the Roman god of war, has a reddish appearance due to the presence of iron oxide (FeO, equivalent to rust) on its surface. It has a thin atmosphere, mostly carbon dioxide, that gives a surface pressure less than one percent that of the Earth. Like the Earth, Mars has ice caps that advance and recede with the seasons. These caps have a permanent component of water ice and a seasonal component of solid carbon dioxide. Its axial tilt is similar to that of the Earth; in the northern summer carbon dioxide vaporizes from the northern icecap and is deposited on the southern icecap and the reverse process happens in the southern summer.

There is strong evidence that Mars once had a considerable amount of water flowing on its surface. Images show what appear to be dried-up river beds and some areas look like the floors of ancient seas. However, in August 2011 there came evidence of the transient occurrence of liquid water on the Martian surface even now (Figure 5.5).

Figure 5.5 Water flows down the side of Newton Crater, seen as dark tendrils in the lower part of the figure (from NASA/JPL — Caltech/Univ. of Arizona).

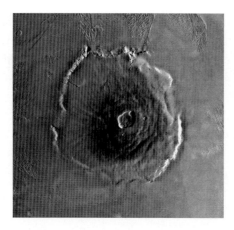

Figure 5.6 A view of Olympus Mons (from Mariner 9, JPL, NASA).

A NASA video showed dark tendrils flowing a few hundred metres downhill from the sides of the Martian crater, *Newton* towards the plain below. The tendrils split on meeting obstructions and then combined on the other side. These events occur in the Martian spring and summer and are thought to be due to the warming effect of the Sun on sub-surface frozen salt water. The presence of liquid water has renewed interest in the search for life on the planet.

Mars has many extinct volcanoes, with the distinction of having the largest volcano in the Solar System, Olympus Mons (Figure 5.6), which rises 24 kilometres above the surrounding plain and is 600 kilometres across its base. Evidence of one-time extensive volcanism on Mars is apparent in the form of solidified lava flows running downhill from the volcanic centres.

Mars shows hemispherical asymmetry with the surface divided into two distinct regions by a scarp (a steep slope) about 2 kilometres high, which runs at an angle of 35° to the equator. North of the scarp is a volcanic plain with few craters; early damage was covered by subsequent flows of lava. These lava flows are estimated to have ceased about 3.8 billion years ago so that the few craters now visible have been produced since then. By contrast, the southern highland region is heavily cratered and shows all the damage inflicted on it over the history of the planet. Figure 5.7 shows the topology, where

Figure 5.7   The topology of Mars.

red indicates the highest regions and, passing through the spectrum, blue the lowest. The deep blue region in the south is the Hellas Basin, a depression 1,800 kilometres in diameter and 3 kilometres deep, caused by the impact of a huge projectile.

The smallest and innermost terrestrial planet, Mercury, is affected by solar tidal forces, a consequence of which is that its spin and orbital periods are related, with the spin period, 58.65 days, being exactly ⅔ of the orbital period, 87.97 days. Because of this relationship, there are two points on the surface on opposite sides of the planet that are directly under the Sun at alternate perihelion passages. Since Mercury has no atmosphere, there are no atmospheric currents to transport heat. At any time the surface temperature varies from 90 K at the furthest point from the Sun to 740 K facing the Sun — a temperature high enough to melt lead, zinc and tin.

Part of the surface of Mercury is shown in Figure 5.8. The edge of the Caloris Basin, a large ringed impact feature some 1,500 kilometres in diameter, can be seen at the left-hand side of the figure. The Caloris Basin is one of the regions directly facing the Sun at alternate perihelion passages. The cratered surface of Mercury shows that it has been heavily bombarded; between the craters are smooth plains suggesting that before the bombardment the surface had been covered by volcanic lava.

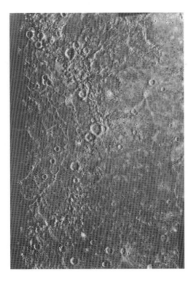

Figure 5.8 The surface of Mercury with part of the Caloris Basin at the left-hand edge (from NASA).

Mercury has a high density, just a little less than that of the Earth. However, the density of the Earth is affected by compression due to the pressure applied by its own mass. Compression is a minor effect on Mercury, so the intrinsic density of Mercury is actually much higher than that of the Earth — 5,300 kg m^{-3} against 4,400 kg m^{-3}. The iron content of Mercury is considerably higher than that of any other terrestrial planet, something that must relate to its formation or evolution. In February 2013, NASA's Messenger probe analysed the surface of Mercury and found materials containing volatile components, such as sulphur and potassium, which would be unexpected if Mercury formed in its present location, so close to the Sun. It has been suggested that Mercury actually formed further out in the Solar System and then moved inwards to its present location.

Chapter 6

Satellites and Rings

They have likewise discovered two lesser stars or satellites, which revolve around Mars, whereof the innermost is distant from the centre of the primary exactly three of his diameters, and the outermost five; the former revolves in the space of ten hours and the latter in twenty-one and a half.

Jonathan Swift (1667–1745) *Gulliver's Travels* (1726)

6.1 General Comments about Satellites

All the major planets have satellite families that were detected by telescopic observation well before the space age. However, the numbers of known satellites for major planets has been greatly expanded by spacecraft observations. The fact that all four major planets have extensive satellite families may be taken to indicate that satellites are produced as a natural concomitant of the formation process of the major planets.

Two of the terrestrial planets have satellites — the Earth has the Moon and Mars has two tiny satellites. The association of satellites with terrestrial planets seems rather more accidental than systematic and this almost certainly indicates some intrinsic difference in the way that major and terrestrial planets formed.

When we come to consider the satellites of individual planets we shall find that some of them are in close-to-circular orbits close to the equatorial plane of the planet. These are referred to as *regular satellites* and it is usually assumed that these satellites were produced as part of the process that produced the planet. We have to be cautious here; spinning planets take on a flattened form with a bulge of material round their equators and it is possible that the gravitational pull of this bulge, plus tidal effects operating on the satellite, could transform it from an irregular orbit to a regular one. Again, the fact

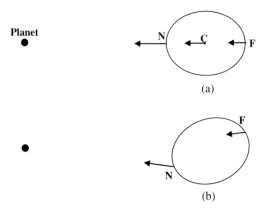

Figure 6.1 (a) The forces on the satellite at the nearside, N, centre, C, and far side, F. (b) The force at N pulls it towards facing the planet. The force at F, which tends to twist the satellite in the other direction, is smaller.

that a satellite orbit does not have the regular form does not mean that it did not have one originally; its original orbit may have been changed by an interaction with some other body, perhaps with one coming in from outside.

Another characteristic shown by most satellites is that they always present the same face to their parent planet; we know that from Earth we always see the same hemisphere of the Moon. This is due to tidal effects. Figure 6.1a shows that the force per unit mass at N, the nearside, is greater than that at C, the centre of the satellite, so it is pulled outwards relative to C to produce a tidal bulge. Similarly the force at C is greater than that at F, the far side, so, relative to C, material at F is also pulled outwards producing another tidal bulge there. Figure 6.1b shows that if the bulge at N departs from facing the planet then the gravitational force of the planet force on it pulls it back. There is an opposite-acting force at F but, since it is more distant from the planet, the force is smaller and it is the one at N that dominates.

6.2 The Satellites of Jupiter

Before the space age there were 12 known satellites of Jupiter, including those discovered by Galileo — Io, Europa, Ganymede and

Table 6.1 Jupiter satellites with average diameter greater than 10 km. *i* is the inclination of the orbit to the planetary equator, *a* the semi-major axis of the orbit and *e* its eccentricity

Satellite	$i(°)$	a $(10^3$ km)	e	Mass $(10^{22}$ kg)	Average diameter (km)	Density $(10^3$ kg m^{-3})
Metis	0.06	128	0.000		40	
Adrastea	0.03	129	0.002		20	
Amalthea	0.37	181	0.003		190	
Thebe	1.08	222	0.018		100	
Io	0.05	422	0.004	8.93	3,630	3.5
Europa	0.47	671	0.009	4.88	3,122	3.0
Ganymede	0.20	1,070	0.001	14.97	5,262	1.9
Callisto	0.20	1,880	0.007	10.68	4,821	1.8
Group of four	25 to 29	11,188 to 11,778	0.13 to 0.21		16 to 170	
Group of four	147 to 164	21,455 to 24,056	0.17 to 0.38		28 to 60	

Callisto. Now we know that there are at least 63 satellites. The ones with mean diameter larger than 10 km are shown in Table 6.1. It will be seen that those out to Callisto seem to be regular while the two outer groups are irregular. Members of the outermost group are in retrograde orbits and these satellites are certainly captured objects.

The Galilean satellites are all different and all interesting. In 1979, an article by three American astronomers, S.J. Peale, P. Cassen and R.T. Reynolds, appeared in the journal, *Nature*, prediciting that Io would show active volcanism. Shortly afterwards the spacecraft Voyager 1 reached Io and the prediction was found to be true. Figure 6.2 shows a volcanic plume from the volcano *Pele* seen by Voyager I. Subsequently, several other volcanoes were discovered on Io.

The basis of the prediction was that the orbital period of Europa, the neighbouring Galilean satellite, is exactly twice that of Io. Thus the two satellites are always closest together at the same point of Io's orbit and the gravitational nudges received by Io always reinforce each other. Io's orbit becomes slightly elliptical so its distance from Jupiter changes as it orbits the planet. Tidal effects

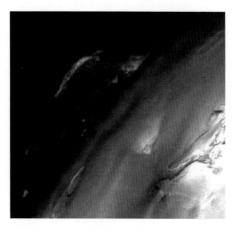

Figure 6.2   The volcanic plume from Pele (from NASA).

tend to stretch a body and the stretch varies from a maximum at perijove[1] to a minimum at apojove. This alternating greater and lesser stretching injects heat energy into Io, an effect similar to what happens when bending a piece of metal backwards and forwards. The bending alternately stretches and compresses different parts of the metal, which produces heat by a process known as *hysteresis*. Because Jupiter is so massive, even the small eccentricity of Io's orbit gives sufficient variation of stretching to provide the energy that drives its volcanoes.

At Io's low temperature, the volcanic material is gaseous sulphur and sulphur compounds. The surface is covered with white and yellow deposits of sulphur dioxide and sulphur (Figure 6.3); the appearance of Io has often been likened to that of a pizza!

Europa has two-thirds the mass of the Moon and is covered with a cracked icy surface. Since there are only three visible craters on its surface, the ice cannot always have been as solid as it is now. At a temperature of about 100 K, water ice is more like solid rock than being the frangible material in a domestic freezer. Because of its interaction with Io, Europa also has a slightly non-circular orbit although the tidal heating within it is much less than for Io since it

---

[1]Perijove is the closest point to Jupiter and apojove the furthest point.

Figure 6.3 The Galilean satellite Io (from NASA).

is further from Juoiter. However, it may be sufficient to give liquid water below the icy surface and it is thought that some form of life might exist in such a sea; to investigate that possibility will be a challenge for some future space mission.

Ganymede, is the largest and most massive satellite in the Solar System with a mass over twice that of the Moon and diameter 10% greater than that of Mercury. However, its low density suggests that its composition is 50% ice and 50% silicate, probably with a small iron core. Some parts of its surface are heavily cratered while other areas show little cratering and are covered by complicated systems of intersecting ridges. The cause of these features is unknown.

The fourth Galilean satellite, Callisto, is slightly smaller than Mercury and has 1.4 times the mass of the Moon. Like Ganymede, its composition is an equal mixture of ice and silicates. Callisto's heavily cratered surface must be very old. There are several multi-ringed features on Callisto similar to the Caloris basin on Mercury. The largest of these is *Valhalla* (Figure 6.4), which was formed by the collision of a very massive projectile.

The three innermost Galilean satellites have a remarkable relationship between their orbital periods, which are almost in the ratio 1:2:4. While these ratios are not precise, there is an *exact* relationship

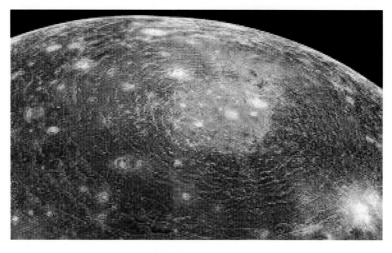

Figure 6.4 The surface of Callisto showing many craters and the ring system Valhalla.

linking their rotational speeds (RS) around the planet. This relationship is:

RS of Io + 2 × RS of Ganymede = 3 × times RS of Europa.

This relationship has been explained by the interplay of gravitational and tidal forces involving the three satellites and Jupiter.

6.3 The Satellites of Saturn

Observations of the space age have increased the number of known satellites of Saturn from nine to 61, although the newly-discovered ones are mostly very small. Those with diameter greater than 20 km are listed in Table 6.2.

The largest satellite, Titan, is second only to Ganymede in both diameter and mass and is the only other satellite larger than Mercury. Its density, 1,900 kg m^{-3}, shows that it must consist of a roughly equal mix of silicate and ice.

Titan has a very dense atmosphere consisting of nitrogen with about 1 percent of methane and traces of other gasses, some of

Table 6.2 Saturn satellites with average diameter greater than 20 km. *i* is the inclination of the orbit to the planetary equator, *a* the semi-major axis of the orbit and *e* its eccentricity

Satellite	*i*(°)	*a* (10³km)	*e*	Mass (10²⁰ kg)	Average diameter (km)	Density (10³ kg m⁻³)
Pan	0.00	134	0.000		30	
Atlas	0.00	138	0.001		31	
Prometheus	0.01	139	0.002		36	
Pandora	0.05	142	0.004		31	
Epimetheus	0.34	151	0.010		113	
Janus	0.17	151	0.007		179	
Mimas	1.57	185	0.020		397	
Enceladus	0.01	238	0.005		504	
Tethys	0.17	295	0.000	7.55	1,066	1.21
Telesto	1.16	295	0.000		24	
Calypso	1.47	295	0.000		31	
Dione	0.00	378	0.002	10.5	1,123	1.43
Helene	0.21	378	0.002		33	
Rhea	0.33	527	0.001	24.9	1,529	1.33
Titan	0.35	1,222	0.029	1,345.20	5,151	1.88
Hyperion	0.57	1,481	0.123		292	
Iapetus	7.57	3,561	0.029	18.8	1,472	1.21
Phoebe	173.1	12,870	0.156		220	
Paaliaq	46.1	15,103	0.363		22	
Albiorix	38.0	16,267	0.477		32	
Siarnaq	45.8	17,777	0.250		40	

which are hydrocarbons. The surface pressure is about 1.6 times that on Earth so that, although it is much smaller than the Earth, it has a greater total mass of atmosphere. Because of the thick atmosphere, visual observation of the surface from spacecraft is not possible but in 2005 the Huygens probe, an ESA component of NASA's Cassini

Figure 6.5 A Huygens image of the surface of Titan. The 'rock' with indicated size is less that 1 m from the camera (from ESA/NASA/PL/University of Arizona).

project, descended to the surface and took photographs, one of which is shown in Figure 6.5. The surface consists of ices of water and various hydrocarbons and the 'rocks' on the surface are mostly small pebble-size pieces of ice.

Several of the other, much smaller, satellites have interesting features. Mimas shows an impact feature that is huge compared to its size (Figure 6.6) and the event that produced this damage must have come close to completely disrupting the satellite.

The satellites of Saturn show several relationships between pairs of orbital periods. The non-adjacent pairs Mimas-Tethys and Enceladus-Dione have a close-to-2:1 ratio of periods, that of Pandora and Mimas is close to 2:3 and that of Hyperion to Titan is 4:3. Another kind of relationship is when a larger satellite has two smaller satellites in the same orbit, one of which leads it in its orbit by 60° while the other trails it by 60°. Thus Tethys has two co-orbiting satellites — Telesto 60° ahead of it and Calypso 60° behind — and, in the same way Dione has Helene leading it by 60° and a small satellite Polydeuces, with

Figure 6.6   Mimas showing the huge impact crater Herschel.

diameter 3 km, trailing behind by 60°. This arrangement of orbiting bodies occurs in another context, described in Section 7.2.

Rhea, the second-largest satellite of Saturn, has an ancient surface, covered by many small craters on the hemisphere facing the direction of motion around Saturn, which makes it lighter in colour than the trailing hemisphere, which is darker and with fewer craters. By contrast, the leading face of Iapetus is much darker than the trailing face; this seems to be due to the leading face being covered by some dark deposit but the reason for this is unknown.

The main interest in Phoebe, which is small and almost certainly a captured body, is its retrograde orbit. It was the first body encountered by the Cassini spacecraft in 2004 and it was found to be covered with huge craters. Bright regions in the crater walls suggest the presence of ice below the surface, although the actual surface seems to contain little, if any, ice.

6.4 The Satellites of Uranus

The five largest satellites of Uranus were known from telescopic observations and are all regular. Spacecraft observations revealed a further 22 satellites; those with diameters greater than 60 km are listed in Table 6.3.

Table 6.3 Uranus satellites with average diameter greater than 60 km. *i* is the inclination of the orbit to the planetary equator, *a* the semi-major axis of the orbit and *e* its eccentricity

Satellite	$i(°)$	a $(10^3$ km)	e	Mass $(10^{18}$ kg)	Average diameter (km)	Density $(10^3$ kg m$^{-3})$
Cressida	0.006	61.8	0.000		80	
Desdemona	0.111	62.7	0.000		64	
Juliet	0.065	64.3	0.001		94	
Portia	0.059	66.1	0.000	1.7	135	
Rosalind	0.279	69.9	0.000		72	
Belinda	0.031	75.3	0.000		90	
Puck	0.319	86.0	0.000	2.9	162	
Miranda	4.232	129.4	0.001	66	472	1.5
Ariel	0.260	191.0	0.000	1,350	1,158	2.1
Umbriel	0.205	266.3	0.000	1,170	1,169	1.7
Titania	0.340	435.9	0.001	3,530	1,578	2.1
Oberon	0.058	583.5	0.001	3,010	1,523	2.0
Caliban	139.9	7,231.0	0.159		72	
Sycorax	152.5	12,179.0	0.522	2.3	150	

The densities of the larger satellites suggest that they consist of mixtures of silicate and ice. The surface of Miranda (Figure 6.7a) is considerably disturbed with regions of parallel ridges and troughs, possibly due to tidal action from Uranus. The surface of Ariel (Figure 6.7b) has been heavily bombarded and shows extended valley systems that may be due to tension in the crust at some stage in its development.

6.5 The Satellites of Neptune

The 13 known satellites of Neptune are listed in Table 6.4. The two known before the space age were both remarkable and certainly not regular. The smaller satellite, Nereid, has the greatest orbital eccentricity of any satellite, 0.749. Its large average distance from Neptune, more than five million kilometres, made it readily visible by telescope observation, although a larger satellite, Proteus, discovered

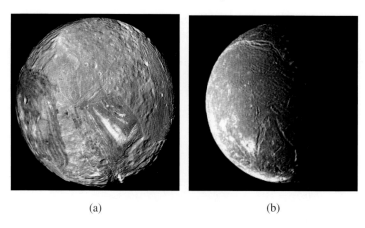

(a) (b)

Figure 6.7 (a) The surface of Miranda showing parallel ridges and troughs. (b) The bombarded surface of Ariel showing fault scarps (from NASA).

Table 6.4 All known Neptune satellites. i is inclination of orbit to planetary equator, a the semi-major axis of the orbit and e its eccentricity

Satellite	$i(°)$	a (10^3 km)	e	Mass (10^{18} kg)	Average diameter (km)	Density (10^3 kg m^{-3})
Naiad	4.7	48.2	0.000		66	
Thalassa	0.2	50.1	0.000		82	
Despina	0.1	52.5	0.000		150	
Galatea	0.1	62.0	0.000		176	
Larissa	0.2	73.5	0.001		194	
Proteus	0.6	117.6	0.000	50	402	
Triton	156.8	354.8	0.000	21,400	2,707	2.05
Nereid	27.6	5,313	0.751	31	340	
Halimede	134.1	15,728	0.571		62	
Sao	48.5	22,422	0.293		44	
Laomedeia	34.7	23,571	0.424		42	
Psamathe	137.4	46,695	0.450		38	
Neso	132.6	48,387	0.495		60	

by Voyager 2 orbiting much closer to Neptune, was previously unobserved.

Before the visit of Voyager 2 it was thought that Triton might be the most massive satellite in the Solar System, but it is actually the

Figure 6.8   Triton showing a frozen icecap from Voyager 2 (NASA).

seventh most massive satellite with about one-seventh the mass of Ganymede. It is in a close retrograde orbit of small eccentricity, substantially inclined to the planet's equatorial plane. This relationship with Neptune is usually taken to imply that it is a captured body.

Triton has a tenuous atmosphere of nitrogen with traces of methane, with a surface pressure about 10^{-5} that on Earth. The surface of Triton has few distinct features, which suggests that it is a comparatively young surface. There is some evidence of flow features and volcanic activity based on volatile materials such as nitrogen, methane and possibly water. A notable feature of Triton is its frozen icecaps, probably of solid nitrogen, seen in Figure 6.8.

6.6 The Earth's Satellite, the Moon

The Moon is very large in relation to its primary body, the Earth. The lunar radius, 1,738 km is more than a quarter that of the Earth and its mass, 7.349×10^{22} kg, is approximately 1/81 of that of the Earth. By virtue of its proximity and comparatively large mass, the Moon exerts an important influence on the Earth though tidal effects.

In general the Moon shows a mixture of bright areas, which are mountainous, and dark areas, which are flat plains (Figure 6.9). The

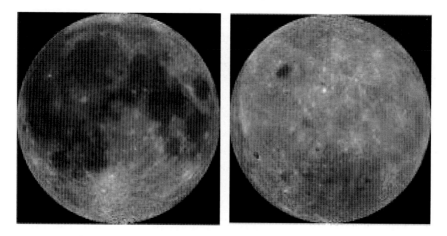

Figure 6.9 The near and far sides of the lunar surface.

latter are termed *maria* (seas; singular *mare*) although most regions of the Moon contain no water, not even bound into minerals. It was assumed that the Moon contained no water at all, but in March 1998 water ice deposits were discovered near both lunar poles by the orbiter *Lunar Prospector*. This water was possibly deposited by comets falling into deep craters near the poles, which were protected from evaporation by being shielded from sunlight. Then in 2009 a 2.2 tonne Centaur rocket was crashed into the 100 km diameter Cabeus Crater near the lunar south pole. The resultant plume of debris, 2 km high, contained copious amounts of water ice and water vapour.

Craters, which are seen all over the Moon, but most densely in the highland regions, are caused by the impact of large projectiles. Such damage is seen on many solar-system bodies. The mare basins were produced by the impact of very large projectiles that produced what were essentially large craters that subsequently filled with molten material from below. The material forming the mare basins is basalt with a composition similar to that from terrestrial volcanic eruptions. As is true for the Earth, the material is a sample of the interior composition at that location. The eruptions producing the mare deposits were spasmodic over a long period of time. Individual flow fronts can be seen in the Imbrium basin, emanating from the vicinity of the crater Euler, and a layered mare was reported by the Apollo 15 astronauts at Hadley Rille, a canyon feature, which had cut through the mare flow to reveal the layers in the local terrain. The general appearance of the layers, plus the extreme flatness of the mare basin, suggest that the lava flows were more fluid than on Earth and so spread further and more evenly.

The time of eruption of maria material has been dated between 3.16 and 3.96 billion years ago which shows that volcanism occurred on the Moon at least during that period but, since older material tends to be covered up, there was probably volcanism even earlier, perhaps back to the origin of the Moon about 4.5 billion years ago.

One of the very early space missions to the Moon, the Luna 3 mission by the USSR in 1959, produced poor pictures of the far side of the Moon but they were good enough to show that the two hemispheres were substantially different. Figure 6.9 shows better quality images of the two hemispheres. While the side seen from Earth is dominated by maria, the far side has few maria, and those relatively small. The hemispherical asymmetry of the Moon is clearly a consequence of some important aspect of either its origin or its evolution.

6.7 The Satellites of Mars

The quotation which heads this chapter, from Jonathan Swift's *Gulliver's Travels*, reports that the scientists of Laputa, an island floating in the sky, had discovered two small satellites of Mars and Swift gave their orbital characteristics. Swift died in 1745 but in 1877 an American astronomer, Asaph Hall (1829–1907), discovered that there *were* two small satellites, close to the planet as the fictional Laputian astronomers had reported — a remarkable coincidence!

The orbital characteristics of the satellites, Phobos and Deimos, are given in Table 6.5, together with the mean diameters. Both satellites are roughly ellipsoidal in shape with dimensions $27 \times 22 \times 19$ km for Phobos and $10 \times 12 \times 16$ km for Deimos. Images of the satellites are shown in Figure 6.10.

Phobos is covered with craters the largest of which, Stickney, has a diameter of 10 km. Phobos orbits Mars in a direct sense but with

Table 6.5   Characteristics of the Martian satellites

Satellite	$i(^{\circ})$	a (km)	e	Average diameter (km)
Phobos	1.0	9,377	0.015	22.2
Deimos	0.0	23,460	0.001	12.6

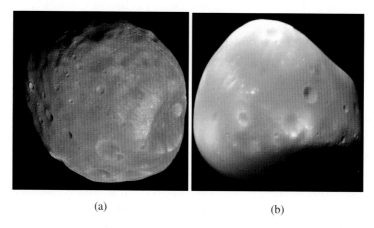

(a)   (b)

Figure 6.10   The satellites of Mars: (a) Phobos, and (b) Deimos.

an orbital period, 7 hours 40 minutes, that is less than the spin period of Mars. As seen from the Martian surface Phobos would rise in the west and set in the east. Both satellites are dust covered, probably from the debris of collisions they have undergone. Despite their orbital characteristics, which seem appropriate for regular satellitess, they are almost certainly captured bodies.

6.8 Ring Systems

The ring system of Saturn, first detected, but not recognized, by Galileo has been studied for four hundred years. A ring system for Uranus was inferred from stellar occultation, i.e. the blocking out of the light from a star by an intervening body, but its exact form was not known at the time of its discovery. Observations from the Voyager 1 and Voyager 2 spacecraft have shown these ring systems in more detail and also revealed that Jupiter and Neptune also possess rings. Thus a ring system accompanies each of the major planets.

6.8.1 *The rings of Saturn*

As revealed by Voyager 1 images, Saturn's rings have a very detailed structure (Figure 6.11). There are two bright rings, an outer ring A and an inner ring B, separated by the Cassini division. Closer in there is a fainter ring C, which is much harder to see and usually referred to as the Crepe ring. It was shown by Clerk Maxwell (1831–1879) in 1857, as an Adams Prize Essay at the University of Cambridge, that the rings could only be stable if they consisted of small solid bodies in independent orbit around Saturn. Modern infrared data shows that they are either icy bodies or perhaps ice-covered silicate bodies. Radar observations indicate that the size of the bodies varies from being small grains up to having diameters of several metres.

The major divisions, already referred to, could be seen by telescopic observation but Figure 6.11 shows numerous finer divisions. These are due to perturbations by various satellites. This is illustrated

Figure 6.11 A false-colour view of Saturn's rings.

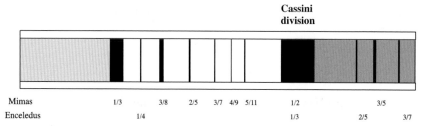

Figure 6.12 A representation of the main divisions in Saturn's rings showing how the periods of bodies within the divisions would relate to the periods of Mimas and Enceladus.

in Figure 6.12, which shows how some of the more distinct divisions, including the Cassini division, are related to the orbital periods of Mimas and Enceladus.

There is uncertainty about the origin of the material of the ring system. One theory is that the orbit of a small satellite decayed until it was so close to Saturn that it was torn apart by tidal forces. However, even if a satellite had been disrupted it would probably have broken up into bodies a few tens of kilometres in diameter that could have subsequently survived; the effectiveness of tidal forces increases with the size of the body on which they act. Subsequent collisions between these initial kilometre-sized bodies could then have produced the finer material to form the rings. Another idea is that two bodies in heliocentric orbits collided close to Saturn and directly produced the small bodies that now constitute the rings.

6.8.2 *The rings of Uranus*

In 1977, when the occultation of a star by Uranus was being observed, it was found that some time before the occultation was due to appear the star was obscured five times by objects close to the planet. After the main occultation by Uranus there was a symmetrical set of smaller occultation on the other side of the planet indicating that they were due to a ring system. Additional occultation measurements increased the number of inferred rings from five to nine. Later, when Uranus was visited by Voyager 2, 11 rings were observed, with the possibility that other very faint rings may be present but unobserved. All the rings are between 42,000 and 51,000 km in radius and seem to consist of metre-sized chunks of ice that are very dark in colour.

6.8.3 *The rings of Jupiter*

In 1979, after the rings of Uranus had been detected, the Voyager 1 spacecraft discovered that Jupiter also had a faint, almost invisible, ring system that can be seen by scattered light at a high angle of incidence. From the infrared spectrum of the scattered light, it is deduced that the particles forming the ring system are small (~3 mm in diameter) silicate grains.

6.8.4 *The rings of Neptune*

When Voyager 2 reached Neptune in 1989 it was found that it too had a set of rings, five in number. Two of the rings are sharply defined while the others are rather faint and tenuous, like the Crepe ring of Saturn. The outer distinct ring is quite lumpy in character with three concentrations of density strung out around it like wide beads on a string.

Chapter 7

Smaller Bodies of the Solar System

It has long been an axiom of mine that little things are infinitely the most important.

Arthur Conan Doyle (1859–1930), *Copper Beeches* (1892)

7.1 The Hunt for a Missing Planet

Prior to the discovery of Uranus by William Herschel in 1781 only the planets out to Saturn were known. In 1766 a German astronomer, Johann Daniel Titius (1729–1796) found that the orbital radii of the planets fitted a simple formula but he did not widely publicize his discovery. In 1772 another German astronomer, Johann Elert Bode (1747–1826), did publicize the formula, without attribution to Titius, and it became known as Bode's Law, illustrated in Table 7.1.

When Uranus was discovered and its orbital radius, 19.19 au, was found to closely fit the Bode's-law value, 19.6 au, the law was regarded by many as a fundamental law and a search was begun by many astronomers to discover the missing planet between Mars and Jupiter. On 1st January 1801, the first day of the 19th century, Giuseppe Piazzi (1746–1826), the Director of the Palermo Observatory in Sicily, discovered a body at a distance of 2.77 au, close to the predicted distance; he named it Ceres after the Roman patron god of Sicily. While Ceres is small, slightly less than 1,000 kilometres in diameter and smaller than many satellites, it satisfactorily filled the gap and gave extra credence to Bode's Law. However, when Neptune was discovered in 1846, its orbital radius, 30.07 au, did not fit the Bode's-law predicted value, 38.4 au, and, thereafter, the law fell somewhat out of favour.

Table 7.1   Bode's law

Planet	Orbital radius (au)	Bode's law
Mercury	0.387	0.4
Venus	0.723	0.7 = 0.4 + 1×0.3
Earth	1.000	1.0 = 0.4 + 2×0.3
Mars	1.524	1.6 = 0.4 + 4×0.3
?		2.8 = 0.4 + 8×0.3
Jupiter	5.203	5.2 = 0.4 + 16×0.3
Saturn	9.539	10.0 = 0.4 + 32×0.3

7.2 Asteroids Aplenty

With the discovery of Ceres the gap in Bode's law was neatly filled but that neatness was not to last long. Within six years, three other bodies, smaller than Ceres but still quite substantial, were found at similar distances from the Sun. These discoveries presaged the discovery of vast numbers of small bodies, down to kilometres in size, and mostly, but not always, situated in the region between Mars and Jupiter. They were designated as *asteroids* and it is now estimated that there are a million or more of them. They mostly have eccentricities less than 0.3 and inclinations less than 25°; although there are some with much higher inclinations it is noteworthy that all asteroids are in direct orbits. A selection of some of the more interesting asteroids is given in Table 7.2.

The second to fourth entries in the table show the large asteroids discovered shortly after the discovery of Ceres. Hygeia and Undina, also quite large, are further out, but still orbit in the region between Mars and Jupiter.

The asteroid Apollo is described as Earth-crossing, that is to say that its orbit straddles the Earth's distance, 1 au, from the Sun. When Apollo was discovered in 1932 it came within 3 million km of the Earth and even closer asteroid passages have been observed. In November 2011 the asteroid 2005 YU55, of diameter 400 metres, came to within 325,000 km of the Earth — closer than the Moon. There is a possibility that one of the 2,000 Earth-crossing asteroids could collide with the Earth; the extinction of dinosaurs, about

Table 7.2   A selection of asteroids

Name	Year of discovery	Semi-major axis (au)	Eccentricity	Inclination (°)	Diameter (km)
Ceres[1]	1801	2.75	0.079	10.6	950
Pallas	1802	2.77	0.237	34.9	608
Juno	1804	2.67	0.257	13.0	250
Vesta	1807	2.58	0.089	7.1	538
Hygeia	1849	3.15	0.100	3.8	450
Undina	1867	3.20	0.072	9.9	250
Eros	1898	1.46	0.223	10.8	20
Hildago	1920	5.81	0.657	42.5	15
Apollo	1932	1.47	0.566	6.4	1.5
Icarus	1949	1.08	0.827	22.9	2
Chiron	1977	13.50	0.378	6.9	unknown

65 million years ago, was probably the result of such a collision. Icarus is also Earth-crossing and has the largest eccentricity of any known asteroid, approaching the Sun to within 0.19 au at perihelion. Earth-crossing asteroids are known collectively as *Apollo asteroids*.

Aten asteroids, with semi-major axes less than 1 au, are Earth-crossing and mostly small but spend most of their time within the Earth's orbit. Other asteroids, like Eros, stay outside the Earth's orbit and are Mars-crossing. Chiron's orbit is mostly between Saturn and Uranus and there are probably many other distant asteroids too small to be observed. The diameter of Chiron is unknown but must be at least 100 km for it to be observed at all.

Other interesting asteroids are the *Trojans* that occur in two clusters in similar orbits to that of Jupiter with one group leading Jupiter by 60° and the other group trailing by 60°, similar to the relationship of Telesto (leading) and Calypso (trailing) to Tethys in its orbit around Saturn (see Section 6.3). The Trojan satellites all oscillate about the precise 60° leading and trailing points and they are quite firmly bound to those positions, as can be confirmed by analysis.

---

[1] Ceres is no longer designated as an asteroid (see Section 7.4).

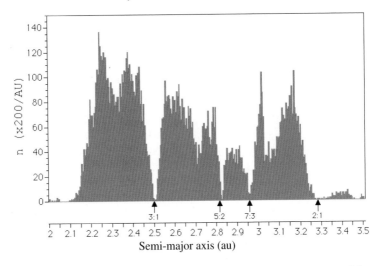

Figure 7.1 An illustration of Kirkwood gaps. The ratios shown with arrows are of the period of Jupiter to that of a body in the gap position (from NASA).

In 1866 the American astronomer, Daniel Kirkwood (1814–1895) noted that when the numbers of asteroids within narrow ranges of semi-major axis are plotted the result showed interesting gaps (Figure 7.1), now known as *Kirkwood Gaps*. Their positions correspond to distances at which an orbiting body would have a period that is a simple fraction of Jupiter's orbital period. The mechanism for this is similar to that which gave the divisions of Saturn's rings, as illustrated in Figure 6.12.

7.3 Meteorites — Chips Off the Old Block

Meteorites are small fragments from asteroids, chipped off by their occasional collisions. Somewhere between 100 and 1,000 tonnes[2] of these asteroid samples fall to Earth each day, mostly in the sea or in inaccessible regions of the Earth, but when they can be recovered they are a valuable source of information about the composition of solar-system material. They arrive in a variety of forms, from small fragments to great boulders weighing many tonnes.

---

[2] The metric tonne is 1,000 kg.

There are three main types of meteorite: *stones*, which are primarily silicates; *irons*, which are two types of iron-nickel alloy; and *stony-irons*, which are mixtures of the two other types of material. For each of these types there are sub-classifications defining fine differences of chemical and physical composition but we shall concentrate on describing the broad characteristics of meteorites, which will suffice for later discussion of how the Solar System may have formed.

In terms of their recovery, meteorites can be divided into *finds* and *falls*. Finds are meteorites that have fallen in the past, perhaps very long ago, but are recognizable as meteorites. In recent years rich hauls of meteorites have been recovered from the Antarctic; the ice there is about three kilometres thick so that one can be reasonably sure that any solid object on or near the surface came from above and not from below. Falls are meteorites that are seen to fall — they become incandescent as they travel at high speed through the atmosphere — and they are collected from the observed area in which they land. It is expected that the proportion of the three types of meteorite that arrive on Earth reflect their proportions in space and that this would be the proportion in falls or in Antarctic specimens. However, stony meteorites, especially when they have been weathered for many thousands of years, can be indistinguishable from terrestrial rocks so the proportion of stones in non-Antarctic finds is expected to be smaller. This conclusion is supported by a 2006 NASA report, the results of which are summarized in Table 7.3.

Table 7.3 The percentages of finds and falls of various types of meteorite from outside Antarctica compared with the percentages from Antarctic finds

Type of meteorite	Non-Antarctic falls	Non-Antarctic finds	Antarctic finds
Stone	93.6	53.4	94.9
Iron	5.1	43.0	4.3
Stony-iron	1.2	3.6	0.9

7.3.1 *Stony meteorites*

The two main types of stony meteorite are *chondrites* and *achondrites*. About 86 percent of stony meteorites are chondrites that usually, but not always, contain *chondrules*, small glassy silicate spheres (Figure 7.2). By contrast achondrites *never* contain chondrules and they resemble terrestrial igneous differentiated rocks (Figure 7.3).

The importance of chondrules is that they show that at some time the temperature was so high that the chondrule material was vaporized. The vapour then cooled quickly and the liquid silicates, condensing out of the vapour, formed small silicate spheres under the influence of surface tension. When they solidified, these spheres were then incorporated into silicate fragments and the material was eventually compressed into rock by gravitational forces. The fact that the chondrules cooled

Figure 7.2 Part of a cross-section of a chondritic meteorite showing chondrules.

Figure 7.3 An achondrite resembling a terrestrial rock.

very quickly is revealed by the composition of the minerals within them. In the vapour phase whole minerals would not have existed but there would have been components such as SiO_2 (silica), Na_2O (sodium oxide), MgO (magnesium oxide) and Al_2O_3 (aluminium oxide) that, combined in various ways, can form different minerals. Combinations of these components produced an initial collection of minerals within the chondrule when it formed and, while the silicate was hot, the components changed partners to form more stable minerals — meaning ones of lower energy. Given enough time the minerals in the chondrule would have had the lowest possible energy consistent with using all available components; the mixture of minerals would then be *equilibrated*. However, the mixtures of minerals in chondrules are *non-equilibrated* — corresponding to a total energy far greater than the minimum possible. This shows that the chondrules became solid and cooled so quickly that the non-equilibrated state became frozen in, since individual components did not have enough energy to jostle their way though the material to form more stable minerals. From the degree of non-equilibration it is deduced that the chondrules cooled quickly, with the rate of cooling in the range 700–4000 K hour^{-1}, suggesting that the total cooling time from the formation of a liquid chondrule to the chemically-frozen state was of the order of one hour.

An important category of chondrites are the dark-coloured *carbonaceous chondrites* (Figure 7.4) that contain volatile materials.

Figure 7.4   Three fragments of carbonaceous chondrites (from NASA).

They can contain up to 22% water, not in the free state but as water of crystallization in minerals, and they also contain organic materials including hydrocarbons such as benzene. Other organic materials include some amino acids, the constituents of proteins, and three of the four basic nucleotide units that form deoxyribonucleic acid (DNA) — the material that gives the blueprints for all forms of life. It has been speculated that life on Earth may have originated from materials brought in by carbonaceous chondrites and comets.

Although they contain so many volatile materials, carbonaceous chondrites also contain some highly refractory substances, materials that are very stable and melt only at very high temperatures. These form white inclusions of minerals; they are called CAI, standing for calcium-aluminium-rich inclusions.

The minority type of stones, the achondrites, have the appearance of igneous rocks that formed on the surface of a planetary-size body. Most of them solidified about the same time — 4,500 million years ago. There are three kinds of achondrite that are different from all others in this respect — the shergottite, nakhlite and chassignite classes, collectively called SNC meteorites — where solidification ages are much more recent, just 1.3 billion years ago in the case of the nakhlites. These meteorites are thought to be material that was chipped off Mars by an asteroid impact. The most compelling evidence to support this hypothesis is that the gas trapped within the meteorites is closely similar in composition to that of the Martian atmosphere.

7.3.2 *Iron meteorites*

Iron meteorites are actually a mixture of iron with from 5–25% of nickel. In the solid state there are two iron-nickel alloys, nickel-poor *kamacite* and nickel-rich *taenite*. Some iron meteorites with 5–6.5% nickel are mostly kamacite with little taenite. Under a microscope they show a characteristic pattern related to the crystal structure of kamacite and they are called *hexahedrites*. At the other end of the scale, with 20% or more of nickel, the meteorite will be almost pure taenite and will show no structure under a microscope; they are called *ataxites*, which means '*without form*' in Greek.

Figure 7.5 Widmanstätten figures.

With between 6.5% and 13% nickel the meteorite will contain a mixture of kamacite and taenite. In the liquid state the two metals were intimately mixed. When it cooled and first became solid it was uniformly taenite. At a certain temperature tiny kamacite plates appeared and nickel migrated out of the kamacite regions into the taenite regions. However, nickel more easily left the kamacite than it entered the taenite so there was a build up of nickel content in a rim around the kamacite plates. With the passage of time the kamacite plates grew but once the material cooled to 800 K, the atoms were frozen in their existing positions and the plates ceased to grow. A pattern called *Widmanstätten figures* (Figure 7.5) can then be seen in a polished plane section of the meteorite. From its appearance, in particular from the size of the kamacite plates, it is possible to estimate the rate of cooling of the meteorite — usually in the range 1–10 K per million years. This is consistent with iron meteorites having been in the deep interior of bodies of small asteroid size or near the surface of larger asteroids.

7.3.3 *Stony-iron meteorites*

Virtually all, meteorites contain both iron and silicate but for about 1% of meteorites both components are present in substantial

Figure 7.6 A Pallasite with olivine globules within a lighter-coloured metal matrix (from Oliver Schwartzbach).

quantities and these are designated as stony-irons. There are two main types of stony-irons — *pallasites* and *mesosiderites*.

For pallasites, silicate materials, predominantly olivine, are set in a metallic matrix (Figure 7.6). This could be derived from a body with a liquid iron core within which a considerable amount of silicate was trapped. Globules of silicate would rise through the molten metal but, as the temperature fell, they would rise at an ever-slower rate until eventually the silicate globs were frozen within a solid metal framework.

Mesosiderites look quite different. The rock is partially in fragments, consisting of minerals, many of igneous origin, that are only stable at low pressures, below 3 kbar (3,000 atmospheres) so they could not have originated from deep within a massive body. The metal is partly in globular form and partly as veins occupying the space between minerals. One way that mesosiderites might have formed is by metallic and silicate materials being violently brought together, which would explain their rather chaotic structures.

7.4 Comets Aplenty

Most people know what a comet looks like, in a very general sense, even if they have never seen one. When they are visible they can be

Figure 7.7 Comet Mrkos (1957) showing a long plasma tail and a stubby dust tail (from Mount Palomar Observatory).

the most awe-inspiring sight that astronomy has to offer (Figure 7.7) and in past times they were thought to be the harbingers of great events — usually catastrophic ones. In his play *Julius Caesar*, William Shakespeare wrote:

> *When beggars die there are no comets seen.*
> *The heavens themselves blaze forth the death of princes.*

A friend of Isaac Newton, Edmund Halley (1656–1742), realised that comets were bodies orbiting the Sun and postulated that the comet seen in 1682 was also the one seen in 1607, 1531 and 1456. He predicted its return in 1756 but did not live to see his prediction confirmed. It is Halley's Comet that is shown in the Bayeaux tapestry, the record of the invasion of England by William I in 1066.

7.4.1 *The structure of comets*

An American astronomer, Fred Whipple (1906–2004) gave the generally-accepted view of the structure of a comet when he described it as a 'dirty snowball'. This means that they are loose agglomerations of ices and silicate dust with an overall density somewhere in the region of 400–700 kg m^{-3}, less than that of water.

Figure 7.8   A spacecraft picture of Halley's Comet (from ESA/NASA).

This 'snowball' is the *nucleus* of the comet, with dimensions of up to a few kilometres. Figure 7.8 shows an image of the nucleus of Halley's Comet taken in 1986, a peanut-shaped object with bright emanations coming from the left-hand side. When the comet approaches the Sun the ices vaporize; the vapour forms a glowing envelope, called the *coma*, with radius between 10^5 and 10^6 km, around the nucleus — considerably bigger than Jupiter. Some of the vapour is dissociated by solar radiation and the hydrogen so produced forms a huge cloud surrounding the coma with ten times its radius. The hydrogen cloud is invisible except to instruments that can detect ultraviolet light.

A feature of a comet, shown in Figure 7.7, is two tails — although for some comets they merge to apparently form a single tail. The longer tail is of gas, made luminous by the excitation of molecules within it by solar radiation, and the stubbier tail is of dust blown off the nucleus when the ice evaporated. It is a popular misconception that the tail points in the opposite direction to that of the comet's motion, like the scarf of the driver of an open-top car. For the comet, the wind that affects the gas and dust is the *solar wind*, due to charged particles moving outwards from the Sun at several hundred kilometres per second. For this reason the gas tail leaves the comet pointing away from the Sun. The dust tail is somewhat less affected by the solar wind and so is slightly displaced from the gas tail.

7.4.2 *The orbits of comets*

There are two categories of comets, *short-period comets* with periods less than 200 years that spend most of their time within the planetary region, and *long-period comets*. Short-period comets with periods more than 20 years can have either direct or retrograde orbits; Halley's Comet, with period about 76 years, has a retrograde orbit. However, comets with periods less than 20 years, called *Jupiter comets*, all have direct orbits, small inclinations, less than 30°, and, for comets, modest eccentricities, usually in the range 0.5–0.7. Their aphelia are all about 5 au, close to Jupiter's orbital radius, and it is thought that they were originally longer period comets that closely interacted with Jupiter and were swung closely round it into their present orbits. A comet loses about one-thousandth of its volatile content at each perihelion passage so that, after about a thousand orbits, taking tens to hundreds of thousands of years, it will lose all its volatile content and thereafter become an inert dark body. Since short-period comets exist now, despite their short active lifetimes, there must be some continuous source of new short-period comets.

There are some very long-period comets, with periods varying from hundreds of thousands to millions of years. Since comets are only seen when they approach close to the Sun, and their periods correspond to very large semi-major axes, they must have aphelia that are tens of thousands of au and eccentricities, close to unity, the limit for an elliptical orbit. This system of comets surrounding the Sun, and stretching out half-way to the nearest star, is called the *Oort Cloud* after Jan Oort (1900–1992), a Dutch astronomer who deduced its existence in 1948. On the basis of some very speculative assumptions Oort estimated that there are about 10^{11} (one hundred thousand million) such comets. Their orbits keep them well clear of the planetary region but occasionally a passing star will nudge a few Oort-cloud comets into orbits taking them close to the Sun and make them visible.

7.5 The Kuiper Belt and Dwarf Planets

In 1951 Gerard Kuiper (1905–1973), a Dutch-born, but later American, astronomer, suggested that outside the orbit of Neptune there should be a population of small bodies in direct orbits around the Sun. His reason for this idea was that he could not accept that the Solar System would simply come to an abrupt end at the distance of the outermost planet. Later it was realised that, if such a population existed, it would solve the conundrum of the source of short-period comets. Bodies in this region would be perturbed, particularly by Neptune, and once in a while one would stray close to Neptune and be thrown into the inner Solar System. In 1992, the first such body outside Neptune, at a distance of 40 au from the Sun, was discovered and since then many hundreds more have been detected. This region of bodies is now known as the *Kuiper Belt* (KB).

Before 2006 a description of the Solar System would have listed nine planets — the ninth member being Pluto, orbiting for the most part outside the orbit of Neptune. At the end of the eighteenth century, when the orbit of Uranus had been reasonably well established, it was found to be wandering slightly from its predicted path. It was suggested that this was due to perturbation by another planet, exterior to Uranus, causing a drift in its predicted path. A young mathematics student at Cambridge, John Couch Adams (1819–1892), became interested in the problem and in 1845, after he graduated, he began to work seriously on the solution, using Newton's law of gravitation. At the end of 1845 he sent his solution, with the predicted position of the new planet, to the Astronomer Royal, George Biddell Airy (1801–1892), but for some reason, still unknown, no search for the new planet was made. Perhaps the work of a new young graduate was not taken seriously enough to justify the use of telescope time.

The problem was independently solved by the French mathematician, Urbain Le Verrier (1811–1877), who specialized in celestial mechanics. On 31st August 1846 he announced his predicted position for the new planet to the French Academy of Sciences and shortly afterwards, using this prediction, it was discovered by Johann Galle (1812–1910) at the Berlin Observatory. Belatedly, the Greenwich Observatory searched using Adams' predicted position and, after the Berlin discovery was announced, they found that they had observed Neptune on two occasions in early August 1846 without recognizing it for what it was.

By 1906 it was mistakenly thought that Uranus and Neptune were both wandering off course and that there could be a ninth planet beyond Neptune. A rich American businessman, Percival Lowell (1855–1916), a keen astronomer, established the Lowell Observatory in Flagstaff, Arizona, in 1894. He searched for the new planet from 1906 until his death but without success. Then, in 1929 a young American astronomer, Clyde Tombaugh (1906 –1997) was given the task of finding the new planet and one year later he was successful. The new planet was given the name Pluto, the first two letters being the initials of Percival Lowell. The eccentric orbit of Pluto kept it mostly outside Neptune but at perihelion it moved slightly inside Neptune's orbit (Figure 7.9). Its high inclination, 17°, plus dynamical effects based on its period being 1.5 times that of

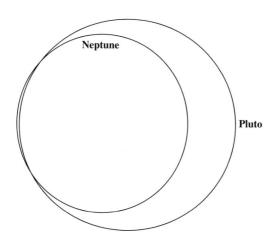

Figure 7.9   The projected orbits of Neptune and Pluto.

Figure 7.10 A Hubble telescope image of Pluto, its large satellite Charon and two small satellites.

Neptune, mean that it never closely approaches Neptune — actually, it approaches closer to Uranus.

The first estimate of Pluto's mass was six times that of Earth but over time this gradually fell to less than one Earth mass. Pluto's planetary status was always disputed but in 1978 it was discovered to have a satellite, Charon, which initially seemed to reinforce its planetary status. Later, a Hubble image (Figure 7.10) showed that Pluto had three satellites, two of them, Nix and Hydra, being very tiny and even later observations have found two further satellites that are even smaller. However, the discovery of Charon enabled a good estimate of Pluto's mass to be found; it was about one fifth that of the Moon, smaller than that of many satellites, and Pluto's planetary status became even more precarious. The final blow came in 2005 when astronomers searching in the KB detected a body, Eris, which also had a satellite (Figure 7.11) and was 27% more massive than Pluto. The astronomical community had a choice — either Eris was a tenth planet or Pluto had to be demoted. At a meeting of the International Astronomical Union in 2006 the latter course was chosen and a new category of bodies was defined — the *dwarf planets* of which both Pluto and Eris were members. To be a dwarf planet a body has to be large enough to have been moulded into a spherical form under self-gravitational forces and to be in an independent

Figure 7.11 The dwarf planet Eris and its satellite, Dysnomia (below, left-centre).

Table 7.4 Characteristics of the dwarf planets

Dwarf planet	a (au)	e	$i(°)$	Diameter (km)	Mass (10^{21} kg)	Satellites
Ceres	2.77	0.080	10.6	975	0.95	0
Pluto	39.48	0.249	17.1	2 306	13.05	3
Haumea	43.34	0.189	28.2	1 150	4.2	2
Makemake	45.79	0.159	29.0	1 500	4?	0
Eris	67.67	0.442	44.2	2 400	16.7	1

orbit around the Sun, the latter part of the definition ruling out satellites. Ceres came into the category of dwarf planets and two others, Haumea and Makemake, have been found in the KB, making five in all — although others might yet be discovered. The dwarf planets and their characteristics are listed in Table 7.4

Chapter 8

The Problem to be Solved

Mankind always sets itself only such problems that it can solve.

Karl Marx (1818–1883), *A Critique of Political Economy* (1859)

8.1 Knowledge and Time

It is sometimes difficult to appreciate the state of knowledge that existed even in the earlier years of our grandparents, forebears that most young people have actual contact with. I knew my great-grandmother, who died when I was fifteen years old and who was born in 1845. The Crimean War ended when she was eleven years old, a war in which more soldiers died of disease than in battle. These were mostly diseases that can now be cured almost routinely by the application of antibiotics. At the beginning of the twenty-first century the great scourges of cancer and AIDS kill millions every year. My grandchildren, who nonchalantly operate mouses (or even mice!) on their computers, will probably look back on my generation as one of pathetic ignorance in which people died of cancer and AIDS because of the lack of some routine treatment. That is the way of the world. Knowledge advances with time and we can only properly judge the achievements of any generation in terms of the knowledge base within which it operates. Newton's contributions to knowledge may seem rather ordinary and straightforward now, but not in the seventeenth century.

What is true for medicine is also true for knowledge about astronomical matters. Uranus was discovered in 1781, the first asteroid (now dwarf planet) was found in 1801, Neptune discovered in 1846 and Pluto in 1930. Any theoretician working on the problem of the origin of the Solar System does so within the framework of the knowledge of his time. He is in the position of a detective faced with the task of solving a crime for which there is no witness. His clues are the observations about the present state of the Solar System and he will try to deduce some sequence of events that will explain them. He must use his imagination, but he cannot let his imagination run riot because there are certain scientific laws that must be obeyed. When he comes up with a proposed solution to the mystery he must present it to a remorseless judge and jury, the scientific community, that will never say that he is right but might, reluctantly, concede that his analysis is plausible — but only that. There is no verdict that stands for all time. If new evidence comes forward that throws doubt on the solution then it will be discarded. One engaged in this area of investigation is in the position that, no matter how many facts his or her theory explains, if it runs counter to just *one* important fact or physical principle then the theory is wrong.

With this picture in mind we first shall look at a range of theories proposed for the origin of the Solar System up to about the year 1960. These theories had very limited objectives — just to explain the existence of the major bodies in the system — and there was no attempt to explain any of the finer detail, such as the presence of asteroids and comets.

8.2 Very Basic Requirements for a Solution

Seen at the coarsest level, the Solar System consists of the Sun and a planetary family, with satellites for the major planets. The planets are all in orbit around the Sun in a *direct* sense (anticlockwise as seen from the north) and their orbits are approximately coplanar — the greatest deviation from the mean plane being Mercury at 7°. The planets are divided into two groups, the terrestrial planets occupying the inner part of the system and the major planets further out.

The Sun spins very slowly, about once in 25 days with its spin axis inclined at 7° to the normal to the mean plane of the planets, as seen in Figure 8.1. The Sun has about 700 times the combined mass of the planets.

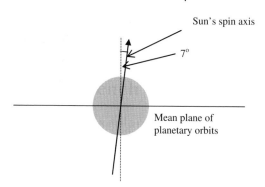

Figure 8.1 The relationship of the Sun's spin axis to the mean plane of planetary orbits.

There are satellite families associated with all the major planets. For Jupiter, Saturn and Uranus the innermost satellites are all in direct near-circular orbits that are almost precisely in the equatorial planes of the spinning planets. Neptune has two unusual satellites, Triton which is in a close *retrograde* circular orbit and Nereid in a very extended and highly eccentric orbit. The satellites associated with the terrestrial planets are somewhat anomalous. The Moon is a very large satellite in relation to the Earth and its relationship to the Earth must be explained by some special event rather than by the normal processes that formed the natural satellites of the Solar System. The two small satellites of Mars are very close to the planet and they have characteristics of size (and appearance — but this was not known in 1960) that are similar to those of asteroids. Again, a special event for their formation is likely.

Given this gross description of the Solar System the following are suggested as essential requirements for any plausible theory of the origin of the Solar System:

1. It must give, or assume, a slowly spinning Sun. This means that if the theory produces the Sun and planets as part of the same process (called *monistic theories*) then the Sun must either be produced spinning slowly or be shown to go through some subsequent

process that slows its spin. However, there are some theories that assume an already-existing slowly-spinning Sun (*dualistic theories*) and concentrate on forming planets. This just sweeps the problem of the slowly spinning Sun under the carpet — but it is still there!

2. It must produce planets in more-or-less coplanar direct orbits. If a theory could give the two different types of planet — terrestrial and major — in the correct general locations then that would be a bonus mark.

3. It should give satellite families for the major planets.

4. It should explain the 7° tilt of the solar spin axis. This is unlikely to be a coincidence since there is only a 1-in-270 probability of having a tilt of 7° or less just by chance.

Of course, much more was known about the Solar System in 1960 than has been described above but for now we ignore this as it was not relevant at that time to the objective of finding a theory of its origin.

Part IV
Early Theories

Chapter 9

The French Connection

The French are wiser than they seem.

Francis Bacon (1561–1626), *Of Seeming Wise* (1612)

9.1 Some Early Theoretical and Observational Developments

The first idea with some kind of scientific basis about the origin of the Solar System was due to René Descartes (1596–1650), a French philosopher and mathematician. At that time, before Newton, although it was known how the planets moved there was no real understanding of the mechanics that governed the behaviour of the Solar System. The Descartes model was rather vague and qualitative and based on observations of fluid motion. Descartes postulated that space was filled with a universal fluid, of unspecified nature, that formed vortices around stars. Eddies in the vortices then produced planets and a smaller system of vortices around planets went on to produce satellite systems. Although it had no physical basis, the theory did deal with the problem of the planarity of the system and the common direction of orbital motion of the planets and satellites. Another idea, that was later developed, was due to Immanuel Kant (1724–1804) who described a process by which a cloud of dust would take on a disk-like form, although the arguments he used would be unacceptable today.

By the end of the eighteenth century there were advances in telescope technology that enabled better observations to be made. William Herschel (1738–1822), the discoverer of Uranus in 1781, the first new planet to be discovered since antiquity, had constructed a magnificent telescope the '40-foot reflector' that was installed in his garden in Bath and was the best of its kind for more than half a

(a) René Descartes   (b) Immanuel Kant

Figure 9.1   Descartes and Kant.

Figure 9.2   William Herschel.

century. With this instrument he observed fuzzy patches of light that he speculated were 'island universes', the first intimation of the existence of other galaxies. However, in 1791 he also observed a nebulosity that seemed to be centred on a single star and in his various writings there can be found suggestions that this was somehow linked to planet formation.

9.2 Laplace and his Spinning Cloud

The theoretical work of Descartes and Kant, and Herschel's observations, were probably influential in the formulation of the first substantial scientifically-based theory of the origin of the Solar System proposed by the French scientist, Pierre-Simon Laplace (1749–1827).

Laplace started his model for the formation of the Solar System with a hot spinning cloud of gas and dust. As it cooled it also collapsed and as it collapsed so it spun faster. This increase in the spin rate is associated with a physical quantity called *angular momentum*, which is so important that it merits a brief description. We consider a body rotating about some axis with a spin rate that, for our purposes, we can think of in terms of 'revolutions per second'. We now think of the body as consisting of a vast number of tiny bodies, that we can call *elements*, each of which has very small linear extent. In Figure 9.4 we schematically show the large body and one of its elements.

Figure 9.3 Pierre-Simon Laplace.

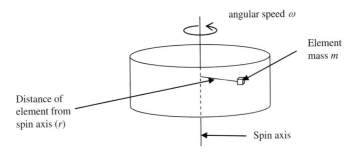

Figure 9.4 An element of a spinning body.

The angular momentum of the element, J, is defined as the product of mass (m) × square of the distance (r) from the spin axis × spin rate (ω), or, in symbols,

$$J = m\, r^2\, \omega. \qquad (9.1)$$

The quantities, added up for all the elements that constitute the body, give its total angular momentum. Angular momentum has the important property that for any body that is completely isolated, i.e. not acted on by forces from outside, the angular momentum remains constant. Actually, this is a matter of common observation. In exhibition ice skating or dancing, the skater will slowly twirl on her skates with arms extended gracefully outwards. She then brings her arms to her side and spins at a much faster rate. What has happened is that by bringing in her arms she has reduced the distance of their mass from the spin axis. To compensate for this and to keep angular momentum constant the spin rate must therefore increase.

Just as for the ice skater, when the spinning cloud shrinks then, to keep its angular momentum constant, it must spin faster and the more it shrinks the faster it spins. As it spins faster so it becomes more flattened along the spin axis until eventually it takes on a lenticular form — the shape of a double-convex glass lens with sharp edges (Figure 9.5).

Eventually the cloud is spinning so fast that material at the edge of the lenticular shape gets left behind in the equatorial plane.

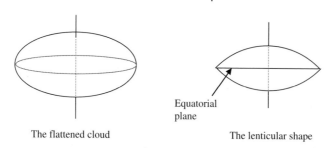

Equatorial
plane

The flattened cloud

The lenticular shape

Figure 9.5   Stages in the collapse of a spinning cloud.

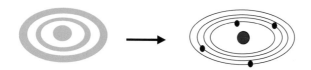

Figure 9.6   The formation of, first, annular gaseous rings and, finally, one planet in each ring.

Laplace postulated that what was left behind would be in the form of a set of rings and that, eventually, the central core of material collapsed inwards to form the Sun. In each ring, material would gradually accumulate by mutual gravitational attraction until it formed a planet. Before the planet formed it would be in the form of a gaseous collapsing sphere and a small scale version of the process that formed planets around the Sun would give satellites around each planet. These final stages of planet formation that, reproduced at a smaller scale also gave satellites, are shown in Figure 9.6.

The theory had an attractive simplicity. A large gaseous sphere, for which there was some observational evidence, spontaneously gave rise to the Sun, planets and satellites by a fairly straightforward mechanical process. Laplace presented his idea in 1796 in a book, *Exposition du Système du Monde*. Initially the theory received great support that lasted throughout the nineteenth century. However, during that period various objections started to be made and gradually doubts began to creep in about the validity of the theory.

9.3 The Problem with a Spinning Cloud

The Laplace model satisfied three of the four basic requirements for a successful theory given in the last chapter. It certainly attempted to explain the existence of the planets and their satellites and gave a planar system, but it made no reference at all to the spin of the Sun. By the middle of the nineteenth century there were many criticisms of Laplace's model but the one that had the greatest impact involved the concept of angular momentum. The criticism was expressed by several individuals in different ways but the nub of the problem is the fact that the Sun, with 99.86% of the total mass of the system, contains in its spin only 0.5% of the total angular momentum of the system. The planets, with 0.14% of the total mass, have 99.5% of the angular momentum contained in their orbits around the Sun. There did not seem to be any reasonable way that the mass and angular momentum of the original nebula could become partitioned like this.

Another expression of the problem is to imagine that all the material of the Solar System (virtually all in the Sun) is expanded to occupy all the space out to Neptune where the first ring was supposed to have detached itself. Since the angular momentum would have to be the same then as it is now (remember, it is constant in an isolated system) then the spin period for the whole cloud would be about 3 million years, much too slow for any ring to have detached itself. Alternatively, if when the Mercury ring detached itself the nebula stretched out to Mercury then it would have been spinning with Mercury's orbital period, 88 days. By the time the remaining material had shrunk to the size of the Sun it would have been spinning with a period of about 17 minutes. In actual fact, the Sun could not spin at that rate because it would fly apart.

There were attempts to rescue the Laplace nebula theory by postulating unlikely distributions of material in the original nebula. For example, a French astronomer, Eduard Roche (1820–1883), considered an original nebula in which most of the material was concentrated at the centre with very little on the outside. This certainly improved the situation with respect to the mass-angular momentum distribution (it was still pretty bad, though) but it meant that there was too little material in the outer part of the nebula to produce planets at all! The various arguments against Laplace's nebula model were so strong that by the beginning of the twentieth century, support for it had virtually ceased and the scientific community was receptive to some new idea.

Chapter 10

American Catherine-Wheels

You're hidden in a cloud of crimson Catherine-wheels.

Christopher Fry (1907–2005), *The Lady's Not for Burning* (1948)

10.1 Spirals in the sky

The American geologist, Thomas Chrowder Chamberlin (1843–1928) examined the Laplace model in some detail, in particular in respect of its implications for the structure and history of the Earth. Eventually he lost faith in the Laplace nebula idea, although not in the idea of the formation of planets from diffuse material. As a result of his background in geology he came to the conclusion that the properties of the Earth would best be explained by an accumulation of solid bodies — which could come about by condensation from some kind of nebula. Early in this work he became associated with a much younger man, Forest Ray Moulton (1872–1952), who was an astronomer and mathematician and whose skills and expertise nicely complemented his own.

In 1900 new observations were being made that were to greatly influence the ideas of Chamberlin and Moulton. These were of spiral nebulae (Figure 10.2) that we now know are complete galaxies, very much like the Milky Way galaxy within which our Solar System resides, but their true nature was then unknown. It was assumed that they were within our own galaxy and each image was interpreted as being a star surrounded by material that could potentially produce planets. Chamberlin and Moulton became convinced that this could provide the scenario for a successful theory of the origin of the Solar System.

Figure 10.1   Thomas Crowder Chamberlin and Forest Ray Moulton.

Figure 10.2   A spiral galaxy.

10.2 Making a Catherine-Wheel

The first idea considered by Moulton and Chamberlin for producing a spiral nebula was as a result of a collision between two neighbouring nebulae. It was soon realised that this would be a very unlikely event and could not plausibly explain the numbers of observed spiral nebulae. Shortly afterwards they concentrated their interest on solar

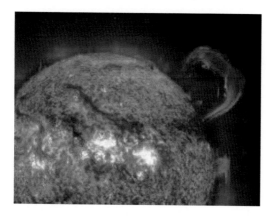

Figure 10.3   A solar prominence.

prominences, large eruptions of matter from the surface of the Sun (Figure 10.3). These eruptions do not lead to a loss of material, as they loop back to rejoin the Sun, but, at the height of a large prominence, material is much more loosely bound to the Sun than when it is part of the surface material.

Chamberlin and Moulton suggested that at a time when the Sun was particularly active a massive star passed close enough to the Sun to pull the solar prominences outwards and leave them in orbit. The tidal effects that caused this to happen would operate most strongly in the direction towards the star and also outwards on the opposite side of the Sun. This is the way tides work. The high tide in the sea caused by the Moon occurs not only at that part of the Earth closest to the Moon but also at the opposite side of the Earth furthest from the Moon.

In Figure 10.4 we see a schematic representation of two streams of matter that have been pulled out of the Sun. The bunching of density in the streams was due to the assumption that the Sun would lose material in a spasmodic fashion. The dense regions would cool rapidly and then liquid and solid objects, that Chamberlin and Moulton called *planetesimals*, would form. Planetesimals would then accumulate to form the planets and smaller collections of planetesimals would go into orbit around the planets to form satellite systems.

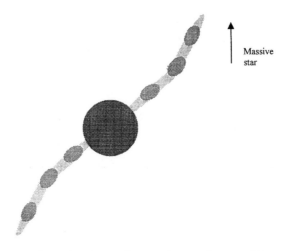

Massive
star

Figure 10.4   Condensations in a solar prominence drawn out by a passing star.

The theory is dualistic in the sense that it assumes a pre-existing slowly rotating Sun before the process of forming the planets begins. However, Chamberlin and Moulton did propose an explanation of the 7° tilt of the solar spin axis as being due to the passage of the massive star that pulled the prominences out of the Sun's equatorial plane. Although the theory has many *ad hoc* features, it does superficially satisfy the basic conditions for a plausible theory.

10.3 Objections to the Chamberlin–Moulton Theory

The Chamberlin–Moulton theory did not receive much support outside the United States, mainly because it was very qualitative with hardly any mathematical analysis to support it. A German astronomer, Friedrich Nölke (1877–1947), put forward some particularly telling arguments against it in 1908. Some of these were:

1. Stellar interactions would be too rare to explain the observed numbers of spiral nebulae.
2. To pull out the prominences, the star would have to be at a distance such that the tidal effects on the two sides of the Sun would be very asymmetrical. By contrast spiral nebulae are quite symmetrical.

3. If the inner part of the spiral corresponded to Mercury and the outer part to Neptune then, because of the very different orbital periods of the planets, the spirals would distort very quickly and the probability of observing them as they are actually seen would be extremely small.

4. Even though the passing star would be fairly close it would not be close enough to give the 7° tilt of the solar spin axis.

All debate about the validity of the Chamberlin–Moulton theory effectively ceased after 1915 once the true nature of spiral nebulae, as whole galaxies, was understood.

Chapter 11

British Big Tides

There is a tide in the affairs of men which, taken at the flood, leads on to fortune...

William Shakespeare (1564–1616), *Julius Caesar*

11.1 The Jeans Tidal Theory

Although by 1915 the Chamberlin–Moulton model had been abandoned, the idea of a tidal interaction between two stars was taken further by the British astrophysicist James Jeans (1877–1946).

The Jeans model, put forward in 1916, was different in that solar prominences were not involved. The basic idea was that a massive star passed close to the Sun and raised huge tides on it — so large that material left the Sun in the form of a filament. This filament then broke up into a series of blobs and each blob eventually collapsed to form a planet. The attraction of the retreating massive star on the blobs pulled them into orbits around the Sun. This interaction is illustrated in Figure 11.2. The blobs, potential planets referred to as *protoplanets*, went into elongated orbits around the Sun. When each blob most closely approached the Sun a smaller-scale version of the same process occurred, in which the Sun pulled out a tidal filament from the protoplanet, thus giving rise to satellite families.

What made this model different from those that preceded it was that Jeans was a brilliant theoretician and he produced mathematical analyses of all aspects of it. He showed that, under the tidal force of the approaching massive star, the Sun would have become distorted into an egg shape and the profile at the small end of the egg would become increasingly sharp as the star approached more closely. Eventually the profile at the sharp end would become a point and thereafter material would escape from it in the form of a stream, or

Figure 11.1 James Jeans.

filament. Jeans then produced further analysis to show that the filament would break up into a series of blobs. The process by which this happens is rather similar to something that is seen when a fine uniform stream of water coming from a tap suddenly breaks up into a string of droplets. For the water stream it is a property of a liquid called surface tension that causes the break-up; for the gas filament it is the force of gravity that gives the instability that creates the blobs. If the material in the gas stream has temperature T (absolute scale) and density ρ in kg m^{-3} (kilograms per cubic metre) then the length, l, of each blob within the stream is given in metres by,

$$l = \sqrt{\frac{\pi k T}{G \rho \mu}} \qquad (11.1)$$

where k is *Boltzmann's constant* 1.38×10^{-23} J kg^{-1} K^{-1} (joules per kilogram per degree absolute), G is the *Gravitational constant*, 6.67×10^{-11} m^3 kg^{-1} s^{-2} (metres cubed per kilogram per second squared) and μ is the mean mass of the molecules that constitute the gas (in kilograms).

Another important theoretical concept introduced by Jeans is the idea of a *Jeans critical mass*. If one imagines a uniform sphere of gas

Forming filament

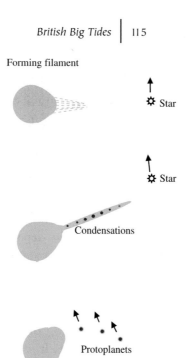

Star

Star

Condensations

Protoplanets

Figure 11.2   The Jeans tidal theory.

with a particular density and temperature then the Jeans critical mass defines the minimum mass for which the sphere would just be able to collapse. The gas sphere is under two influences — gravity that is tending to cause it to collapse and gas pressure that tends to expand it. The force of gravity will depend on the size, and hence the mass, of the sphere — gravity forces are stronger on Earth than on Mercury, although they have similar average densities. We can think of gas pressure as the outward force that keeps a balloon inflated. If we keep the temperature constant but insert more gas then the balloon slightly expands, showing that the pressure became greater as the density of the gas becomes greater. Alternatively, we can keep the content of the balloon constant and place it in hot water to heat up the gas. Again the balloon expands, showing that the pressure has increased despite the fact that the density of the gas will have slightly fallen because the mass of gas has remained constant but is in a slightly greater volume. We can see that pressure of a gas depends on its density and temperature. However, it does not depend on the total amount of the gas present.

If the mass of the gas sphere is smaller than the critical mass then the pressure force dominates and the sphere will expand; if the mass is larger than the critical mass then the gravity force dominates and the sphere will collapse. The critical mass for a sphere can be derived theoretically as,

$$M_J = 5.46\sqrt{\frac{k^3 T^3}{G^3 \mu^3 \rho}}. \qquad (11.2)$$

For an irregular shape or a non-uniform density the numerical constant in (11.2) will be different but the dependence on the properties of the gas would be as indicated. If the mass of a blob in the tidal filament exceeded the appropriate Jeans critical mass for its shape and mass distribution then it would be able to collapse to form a planet.

This theory captured the imagination of the astronomical community and it had very wide acceptance. It was a dualistic theory so it did not have to explain the slow spin of the Sun. It explained the formation of planets, all in direct orbits, and satellites for those planets. Jeans also pointed out that the filament would have a cigar shape and be thickest in the middle when the massive star was closest to the Sun so pulling out material at a greater rate. This would explain why Jupiter and Saturn, the most massive planets, were in the centre of the system. There was no direct explanation of the 7° tilt of the solar spin axis but the spin axis would certainly not be expected to be normal to the plane of the planetary orbits. Those orbits would be close to the plane defined by the motion of the massive star with respect to the Sun but with a component out of that plane because of the Sun's spin. Although there were some loose ends to be tidied up it seemed that, at last, a viable theory for the origin of the Solar System had been produced.

11.2 Jeffreys' Objections

The British geophysicist Harold Jeffreys (1891–1989) was at first a strong supporter of the Jeans model, so much so that it was sometimes referred to as the Jeans–Jeffreys theory, although the two

Figure 11.3 Harold Jeffreys.

individuals never worked together. Nevertheless, in 1929 Jeffreys was the first to express doubts about the idea. The first objection he raised was that the probability of a massive star passing close to the Sun was extremely small. At the time this was not a valid argument. There is a tenet, known as the *anthropic principle*, which is generally accepted, that states that astronomical theories are constrained by the necessity of allowing human existence. That humankind exists cannot be disputed. If it is absolutely necessary for some highly improbable event to have taken place for humans to exist then that event *did* take place. In any case, as far as was known at the time, the Solar System might have been the only planetary system, or one of few, in the Universe so it would then have been rare and potentially the product of a rare event.

The second of Jeffreys' arguments was based on some fairly subtle physics involving a mathematical quantity called *circulation*. What it amounted to was the statement that if the material for Jupiter was drawn out of the Sun then, because Jupiter and the Sun have virtually the same density, they should be spinning at the same rate. In fact Jupiter spins about sixty times as fast as the Sun. Once again this turns out not to be a strong argument since it assumes that the Jupiter material came out of the Sun with the density it has now.

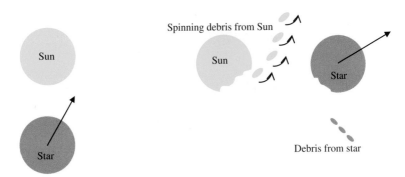

Figure 11.4 Jeffreys' star-collision theory.

If it came out at a lower density, and then contracted, its faster spin can be explained.

In order to deal with this second argument Jeffreys suggested that, instead of a tidal interaction, the star physically sideswiped the Sun so imparting spin to the material that was knocked off to form planets (Figure 11.4). Jeffreys showed inconsistency here since physical contact would be much more unlikely than the tidal interaction that he had criticized on the basis of its small probability. In fact, Jeffreys had returned to a model similar to that proposed in 1745 by the French astronomer Georges-Louis Leclerc, Compte de Buffon (1707–1788). He proposed that a comet had grazed the Sun knocking off material that travelled out to various distances to form the planets. Buffon had no idea of the nature of comets and we know now that their masses are far too small to have much effect on the Sun.

Jeffreys idea of a collision between a star and the Sun did attract considerable support and for some time was preferred to the Jeans model.

11.3 Russell's Objection

The next objection raised against the Jeans theory, and Jeffreys' modification of it, was made in 1935 by the American astronomer Henry Norris Russell (1877–1957). The elliptical orbit of one body around another is closed, meaning that it repeatedly follows the same path in space. Hence if material is pulled out of the Sun into an

Figure 11.5   Henry Norris Russell.

elliptical orbit, then its path will take it back to the Sun and it will be reabsorbed. In actual fact, Jeans model was a little more complicated than this because a blob formed in the filament was affected by the gravitational attractions of the other blobs and of the massive star. Even taking these influences into account, Russell was able to show that material could not be pulled out of the Sun far enough to explain even the innermost planet Mercury, let alone those much further out.

11.4 Spitzer's Objection

The American astrophysicist Lyman Spitzer (1914–1997) put forward an argument against both the Jeans and Jeffereys models based on theory produced by Jeans himself! This involves the *Jeans critical mass* that has already been referred to in respect of blobs collapsing to form planets. Spitzer argued that if the Sun had been in its present condition then an amount of material taken from it sufficient to produce Jupiter would have about the present density of Jupiter, or perhaps less, and a temperature of one million degrees kelvin.

Figure 11.6 Lyman Spitzer.

Applying the Jeans critical-mass formula showed that the pressure force, tending to give expansion, would overwhelm the gravity force and the mass of gas would not stay together to form Jupiter, but would violently dissipate into space.

In summary Russell's and Spitzer's arguments amounted to saying that, not only could planets not form at sufficient distance from the Sun but they could not form at all!

11.5 A Later Objection

Jeans accepted the validity of the theoretical arguments made against his theory and wrote, in a truly objective spirit, "The theory is beset with difficulties and in some respects appears to be definitely unsatisfactory". At the end of this episode one had a model that could not be upheld but one also had a body of theoretical analysis developed by Jeans that was sound and could, perhaps, be deployed in theories yet to come.

Much later, long after the Jeans model had been abandoned, an objection was raised that would apply to both the Jeans model and to Jeffreys' modification of it. The temperature inside the Sun is high enough for many nuclear reactions to take place that involve atoms of low mass. Amongst these 'light atoms' are lithium, beryllium and boron that are transformed into other atoms at solar temperatures. By looking at the Sun's spectrum we can tell what sort of atoms are present and these light elements are conspicuously absent. However, on Earth these elements, whilst not abundant, are present in substantial quantities. The implication is that the material of the Earth did not come from a source as hot as the Sun — and presumably that restriction would apply to all the planets. This objection indicates that we must only consider theories for which planetary material has a 'cold' origin where the term 'cold' in this context can include temperatures up to hundreds of thousands of degrees, but not temperatures at which nuclear reactions would take place.

Chapter 12

Russian Cloud Capture — With British Help

Stooping through a fleecy cloud.

John Milton (1608–1674)

12.1 The Schmidt Model

The demise of the two dualistic theories that had been dominant for the first three or four decades of the twentieth century did not see an end to such theories. The next form of dualistic theory, the Accretion Theory, of a new and interesting kind, came from a Russian planetary scientist, Otto Schmidt (1891–1956) in 1944. Telescope observations show regions of the sky where stars cannot be seen because the light coming from them is absorbed by intervening dense cool clouds of dusty gas. Schmidt reasoned that from time-to-time a star on its journey through the galaxy would pass through one of these clouds and, in doing so, might pick up a surrounding envelope of gas and dust. If it had some rotational motion the envelope would settle down into a disk-like form and from its material the planets would form.

Schmidt assumed that if only the Sun and the cloud were present then capture of cloud material could not take place. This assumption was based on the principle that if two bodies approach each other from an infinite distance apart and do not collide then they must end up an infinite distance apart — which is certainly true for two non-colliding point masses but hardly applies to a star passing through a very extended cloud. For this mistaken reason Schmidt postulated that there had to be another star somewhere in the vicinity, the gravitational field of which would enable the capture of cloud material to occur by removing some energy from the interacting star-cloud system. This requirement made the likelihood of the event very small

Figure 12.1   Otto Schmidt and Ray Lyttleton.

Figure 12.2   The Sun picking up a dusty gaseous envelope after passing through a dense cool cloud.

although, appealing once again to the anthropic principle, that did not necessarily rule out the idea at the time it was put forward. In 1960 the British astronomer Ray Lyttleton (1911–1995) took up Schmidt's idea and showed that it was not necessary to have another star present and that the Sun, after passing through the cloud, could be accompanied by captured material (Figure 12.2).

12.2 Lyttleton's Modification of the Accretion Theory

Lyttleton used theory developed by Hermann Bondi (1919–2005) and Fred Hoyle (1915–2001) that dealt with how a star could capture material from a cloud. Figure 12.3 shows the motion of the cloud material, at speed V relative to the star. The cloud material is attracted towards the axis by the gravitational attraction of the star; the figure shows two streams coming in from distance R on opposite sides of the axis.

When these gas streams interact on the axis they mutually destroy their components of motion perpendicular to the axis, which

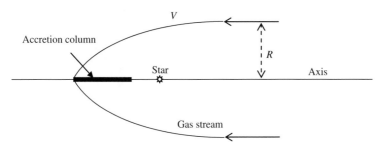

Figure 12.3   Gas streams interacting on the axis.

slows the material down to less than the escape speed from the star. This can happen for pairs of streams coming from up to some maximum distance from all round the axis and the effect is to create a high-density accretion column of material on the axis that is captured by the star. Some of it may impinge on the star but, if there is some residual tangential motion of the colliding streams, then this will lead to the capture of some of it as a surrounding envelope of material.

The original Lyttleton model was unsatisfactory in that it postulated unrealistic conditions — for example, it assumed that the Sun would enter the cloud with a speed of 0.2 km s^{-1}, which is impossibly low since the gravitational attraction of the cloud on the Sun would alone have given a greater relative speed than that. However, in 1973 Chris Aust and I showed that some modifications of the original conditions specified by Lyttleton could lead to the Sun being surrounded with about the right total mass of material at about the right distances to produce the planets. What was missing was a mechanism for turning this very diffuse cloud material into planets — a problem that occurs with another, more modern, theory that will be described in Chapter 21.

In view of its rather vague nature, the theory did not enjoy much support — but the idea of material being captured by the Sun was to return.

Chapter 13

German Vortices — With a Little French Help

'Mind and matter' said the lady in the wig *'glide swift into the vortex of immensity'.*

Charles Dickens (1812–1870), *Martin Chuzzlewit*

13.1 First Ideas About Vortices

When a spoon is moved through a cup of coffee the liquid surface moves around in a set of swirling motions. Larger-scale motions, described as *vortices*, are usually accompanied by smaller swirling motions, known as *eddies*. Such motions are also seen when a fluid moves past a stationary object, such as a post set in a flowing river. We have already referred, in Section 9.1, to the work of René Descartes, the French philosopher, mathematician and physicist, who was a contemporary of Galileo and Kepler and who, like them, believed in the essential correctness of the Copernican model. Descartes' theory dealt with the heliocentric nature of the Solar System, the planarity of the system and the direct orbits of the planets and satellites. Descartes described this idea in a book, *Le Monde*, but, mindful of the way that the Church had treated Galileo, he was fearful to publish it and it did not appear until 1664, after his death.

13.2 The von Weizsäcker Vortex Theory

In 1944, the German astrophysicist Carl von Weizsäcker (1912–2007) considered how a pattern of vortices might be set up in gaseous disk due to turbulence, random motions of a fluid that stir it up in the way that a spoon stirs up liquid in a cup. He suggested that a pattern of vortices was set up as illustrated in Figure 13.2. He

Figure 13.1 Carl von Weizsäcker.

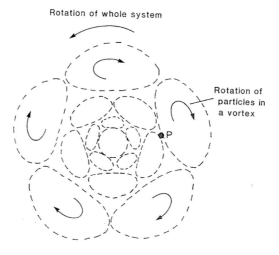

Figure 13.2 The pattern of vortices according to the von Weizsäcker model.

showed that the combination of a clockwise rotation within each vortex together with an anticlockwise rotation of the whole system could lead to each individual particle of fluid moving in an elliptical orbit around the central mass, just as Kepler's first law requires. At a point such as P, where neighbouring vortices met, material would be colliding at high speed and material would coalesce there to form condensations. All the condensations in one ring would eventually combine to form a planet. Von Weizsäcker showed that if there were five vortices to a ring then the pattern of orbital radii would be similar to that observed in the Solar System. As with all evolving disk models the central part of the disk was assumed to eventually form the Sun.

13.3 Objections to the Vortex Idea

The harshest critic of this model was Harold Jeffreys. Von Weizsäcker had appealed to turbulence as the agency for enabling the pattern of vortices to be established. Turbulence causes a loss of energy of a system since colliding turbulent streams of matter generate heat and this heat is then radiated away. The heat energy is derived from the mechanical energy and hence the mechanical energy of a turbulent system gets less. It is a basic physical law that all systems are most stable when they have the lowest possible energy and hence systems move to lower energy states when the conditions allow. Jeffreys showed that von Weizsäcker's system of vortices was a *high* energy system and hence would not form as a result of turbulence. The natural end state for a turbulent system is one of lowest energy and for a disk of the type postulated by von Weizsäcker this would lead to all parts of it moving in smooth circular motion around the central mass.

The proposed mechanism deals neither with the problem of the formation of a slowly spinning Sun nor with the formation of satellites, both of which are basic requirements for any plausible theory.

Chapter 14

McCrea's Floccules

...he made the stars also.

Genesis

14.1 Producing Stars and Planets Together

The general idea behind some of the previously described theories is that an individual mass of material undergoes processes that eventually produce the Sun accompanied by the planets. Observations show that many stars occur in clusters and, indeed, it is generally thought that all sun-like stars originally formed in clusters. This will be discussed further in Chapter 27. In 1960 the British astronomer, William McCrea (1904–1999), took a rather broader view of star and planet formation than hitherto and considered a scenario that would produce a whole cluster of stars together with planets for some of them. He tried simultaneously to solve the problem of why it is that solar-type stars spin slowly and how it is that they acquire planetary systems.

McCrea's starting point was a cloud of gas, mostly hydrogen and helium, with about 1% of its mass in the form of dust. This cloud, which eventually forms a stellar cluster, is in a turbulent state so that individual regions of the cloud are moving in random directions with supersonic speeds. We shall see in Section 27.4 that there is strong observational evidence for turbulent motions in star-forming clouds. When two of these turbulent regions collide, the gas in them is compressed and McCrea modelled the cloud as a large number of high-density regions, called *floccules*, moving in random fashion in a lower-density background. In the first version of the theory the masses of the floccules were about three times the mass of the Earth. When floccules collided they combined and the larger-mass body so

Figure 14.1   William McCrea.

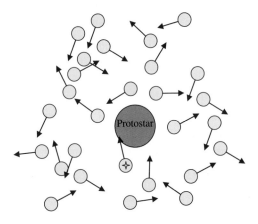

Figure 14.2   Two-dimensional representation of floccules and a growing protostar.

formed is then more able to attract more floccules, thus further enhancing its growth. In this way, regions of the cloud developed in which one dominant body formed as a *protostar*, a massive low-density body that would eventually collapse to become a star. Smaller clusters of floccules became incipient planets, *protoplanets*, which were captured in orbit around the protostars. A schematic representation of this model is shown in Figure 14.2.

A good feature of the model is that it leads to a slowly-spinning Sun. Every time a floccule joined the growing protostar it contributed to its spin according to its speed and the angle and position of its impact; the floccule marked with a cross in the two-dimensional figure contributed to the spin in a clockwise direction. In the three-dimensional situation the contributed spins would have been in all directions at random and they would have tended to cancel out. McCrea showed mathematically that the expected resultant spin of the Sun would have been very close to what is observed.

Another prediction of the model is that the initial protoplanets would have been more massive than the present planets and that they would have spun very rapidly. The fast spin comes about because it took fewer floccules to form a planet so the cancelling-out process for the spins would have been less efficient. This apparent difficulty of the model was interpreted by McCrea as a positive feature. As the protoplanet collapsed so it would have spun ever faster until eventually it became rotationally unstable and broke up. The form of break-up in such a situation had already been investigated by Lyttleton in considering another theory. The protoplanet would disrupt into two parts with one part having about eight times the mass of the other. As the two parts separated so a fine neck, or filament of material, connected them and little blobs forming in this (as Jeans described) would be retained in orbit by the larger part (Figure 14.3). These constituted a satellite family. The smaller portion would have moved at a much greater speed than the larger one and its speed would have been sufficient for it to have escaped from the Solar System. One is left with a slower-spinning protoplanet accompanied by a satellite family, just what is required for the major planets.

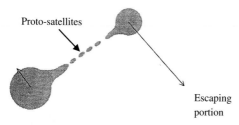

Proto-satellites

Escaping portion

Figure 14.3 The break-up of a rapidly spinning protoplanet.

McCrea suggested that, in the inner part of the Solar System, the process of disruption of a protoplanet into two parts took place involving just the dust component of the material so that two solid terrestrial-type planets were formed. He suggested Venus-Mercury and Earth-Mars as the two pairs produced. In the inner part of the Solar System the speed required to escape is high so that both parts were retained.

By assuming that all the floccules in a particular region of the whole cloud were bound just to the nearest protostar, McCrea showed that the expected number of planets and their orbital radii would be in reasonable agreement with the present Solar System. The theory was a bit lacking in detail and was not computationally simulated but the numerical results were very satisfactory.

14.2 Objections to the Floccule Theory

The first objections to the floccule theory were raised by me in a private communication to McCrea. The theory developed by McCrea gave the density and temperature of the floccules and it turned out that the proposed floccule masses — three times that of the Earth — were far less than the Jeans critical mass. Thus the floccules would have expanded and would have completely dispersed on a timescale of about a year. This would not matter if several floccules could get together to make a larger mass in a short time but, again using McCrea's own figures, the average time for two floccules to get together is about thirty years. McCrea countered this objection by increasing the masses of the floccules to about the mass of Saturn, which solved the floccule instability problem but introduced others. For example, with fewer floccules forming the Sun the process by which the spins contributed by individual floccules cancelled out would have been less efficient and the final spin rate of the Sun would have been much higher. There is also the difficulty that the assumption that floccules of planetary mass are formed is suspect; in Section 27.5, an analysis is described that indicated that the first condensations produced in a cloud are of stellar rather than of planetary mass.

Another problem was McCrea's assumption that each floccule was bound to the nearest protostar and to no other. With this assumption he was able to show that the total angular momentum of the floccules within a region dominated by a particular protostar would be similar to the total orbital angular momentum of the solar-system planets. However, in practice the individual floccules would have travelled throughout the cloud, through regions dominated by different protostars, and a great deal of their energy and angular momentum would have gone into the relative motion of the protostars rather than into the motion of protoplanets around protostars.

Finally we should note that, according to this model, the protoplanets could have been captured in orbits with any orientation so that an initially planar system of planets would not be expected and planets could be orbiting in either a direct or retrograde sense. There may have been some mechanism that pulled all the planets into one plane — for example, if the protostar were surrounded by a disk. Then, if there were fewer retrograde planets, they could have been removed by collisions with some that were in direct orbits leaving the surviving protoplanets in direct orbits to form the present system. Finally, the model gives no obvious explanation of the 7° tilt of the solar spin axis that, for this model, is improbably small — although not impossible for it to give.

Although the theory has too many difficulties to be considered as plausible it does introduce the idea that planetary formation is somehow connected with the formation of a stellar cluster. Previous theories had just considered processes acting on a single cloud of material that would produce one star and its accompanying planets. The idea that planet formation is connected with a stellar cluster will occur again in Chapter 29.

Chapter 15

What Early Theories Indicate

If men could learn from history, what lessons it might teach us!

Samuel Taylor Coleridge (1772–1834)

15.1 Angular Momentum Difficulties

Angular momentum problems occur in two different aspects in the theories that have been described so far. The main repositories of angular momentum in the Solar System are the planets in their orbits — accounting for all but a tiny part of the total angular momentum — and the Sun's spin. Other sources, such as the spin of the planets and the orbits and spins of the satellites, are so tiny by comparison that they can be ignored.

The monistic theories, that start with a single mass of material that is to form the Sun and the planets, have the great difficulty that angular momentum has somehow to be removed from the bulk of the material, which will form the Sun, and be transferred to the small amount of material that will become planets. Neither the Laplace nebula theory nor the von Weizsäcker vortex theory addressed the issue of a slowly rotating Sun. We shall see later (Chapter 19) that the need to solve this problem has led to new ideas about transferring angular momentum in the context of a modern nebula theory.

For dualistic theories the problem of the slow spin of the Sun is avoided — but it could be argued that a complete description of the origin of the Solar System should include an explanation for this important property of its dominant member. However, there are those that take the view that observations show that most Sun-like stars spin slowly and that this is a legitimate starting point for a theory. If this is accepted then the emphasis is just to explain the presence of the planets. But, even with this let-out, the angular momentum problem is not trivially solved. An important problem with the Jeans model, and also with Jeffreys modification, is that material cannot be removed sufficiently far from the Sun to give the existing planets, either by the tidal action of a massive star or by a stellar collision. This is an angular momentum problem of a different kind, where not enough of it is possessed by the planets. The dualistic Schmidt–Lyttleton theory of cloud capture and McCrea's floccule theory both get closer to solving the angular momentum problem. For cloud-capture the captured material can possess the right amount of mass and angular momentum to explain the planets. The situation with respect to the floccule model is less clear-cut since some of the underlying assumptions made by McCrea are suspect and the mechanism has never been properly tested by any kind of modelling procedure.

15.2 Planet Formation

While having material at the right distance from the Sun is a *necessary* condition for a plausible theory that, by itself, is not *sufficient*. It must also be shown that the material forms planets.

None of the monistic theories we have described so far has even considered this problem in any detail. Laplace suggested that clumps in his rings would form by gravitational attraction and that eventually the clumps would combine into a single planet. It is possible to show that, unless his rings had masses very much greater that that of planets, the rings would have been very unstable and would have dispersed to give a disk without rings in a very short time — much shorter than the time required for clumping to take place. The end result would be a fairly structureless disk within which the planets must form — a similar situation to that obtained with the cloud-capture model.

The floccule theory produces planets by concentrating cloud material through collisions. It is certainly true that colliding material would be compressed but it would not produce planetary masses in a large cloud. The turbulent streams in such a cloud would have had masses similar to the Jeans critical mass for the cloud material and these would have been of stellar mass. When they collided, stellar-mass condensations would have been produced.

The only early theory that hints at how planets could be successfully formed is that of Jeans. The break up of a filament into a set of blobs under gravitational effects is well founded theoretically and, as will be shown in Chapter 29, has also been successfully modelled. The problem with the Jeans theory was not that the mechanism for producing planets was unsatisfactory but rather that it was being applied to the wrong material. It is necessary to have material in a filament at a density and temperature that would give planetary mass blobs with greater than the Jeans critical mass. For the densities one might expect in a filament, this requires that the material should be at a temperature much lower than that of typical solar material (Section 11.4), a requirement that is also indicated by the quantities of the light elements lithium, beryllium and boron in the Earth's crust (Section 11.5).

15.3 Indicated Requirements for a Successful Theory

The older theories do not clearly indicate the exact nature of a successful theory but they do offer some possible alternatives. The best indications are for a dualistic theory in which the formation of the Sun is not directly linked to the formation of the planets. However, it really should be necessary, as part of the dualistic approach, to explain the formation of the Sun, and by implication other stars, to have an entirely satisfactory theory of the origin of the Solar System. A dualistic approach has the advantage that it reduces the problem of the formation of the Sun to that of producing a solar mass with solar angular momentum, without the additional complication of *simultaneously* having to produce a much smaller mass, the planets, with much greater angular momentum.

For planet formation cold material is absolutely essential. For this reason it is not possible to have planetary material derived from the existing Sun. This requirement is satisfied by a monistic theory in which both the Sun and the planets were to be formed from cold material. If the material has the density and temperature that enabled a planetary mass of it to satisfy the Jeans mass criterion then, clearly, this would be an advantage. Otherwise a process is required by which diffuse material would be concentrated in some way to form planets.

The planetary material must either originally have, or somehow acquire, the angular momentum required to explain the final planetary orbits. If the potential planetary material and the Sun approached each other in such a way that the material initially had the required angular momentum relative to the Sun then this would be helpful. This general condition is satisfied, more-or-less, by the Schmidt–Lyttleton theory and the McCrea floccule theory.

Bearing these indications in mind we shall be looking at some modern theories to see how well these requirements are satisfied. However, before doing so we shall learn about some recent observations that, in a positive sense, give helpful indications of the kinds of theories that could be successful but, in a negative sense, impose new constraints on theories that they must satisfy.

Part V
New Knowledge

Chapter 16

Disks Around New Stars

Look at the stars! Look, look up at the skies.

Gerard Manley Hopkins (1844–1889),
The Starlight Night

16.1 How Hot and How Luminous?

An important property of any star is its temperature and there is an associated property, its *luminosity,* which is a measure of the rate at which it emits energy by radiation. The radiation it emits is *electromagnetic radiation*, a coordinated electrical and magnetic disturbance propagated in the form of a wave. The wavelength dictates what kind of radiation it is and how it may be detected. For example, if the radiation has a wavelength between 0.4 microns[1] and 0.7 microns then the radiation is visible light with blue and red at the short and long wavelength ends, respectively. Going from blue towards shorter wavelengths first traverses the ultra-violet (UV) region, then X-rays and finally very short wavelength γ rays with wavelengths from one ten thousandth to one ten millionth of that of visible light. Starting from the red end and going towards longer wavelengths first goes through the infrared region which gradually merges into radio waves with wavelengths up to 10^{11} (one hundred thousand million) times the wavelengths of visible light. This range of electromagnetic radiation is illustrated in Figure 16.1; it is interesting to note how little of it we can detect with our eyes!

It is a matter of everyday experience that when bodies are heated to a high enough temperature they emit light and that the colour of the light changes with temperature. A piece of iron heated to a

---

[1] one micron (μm) = one millionth of a metre.

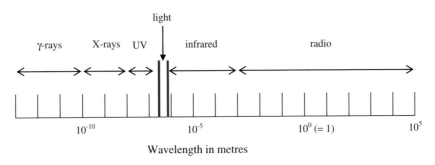

Figure 16.1   The electromagnetic spectrum.

temperature too low to emit light (i.e. an iron for pressing clothes) will still emit radiation in the form of heat, which we can think of as in the infrared region. Heated the iron further, as would be done by a blacksmith, gives visible light that at first is a very dull red but becomes a brighter red as the temperature increases. As the temperature increases further, as for the tungsten filament of a lamp, so the light changes first to a yellow colour and then becomes a brilliant white. At an even higher temperature the white light becomes tinged with blue. As it is for a piece of iron or a tungsten filament, so it is for a star. Lower temperature stars shine with a reddish light while high temperature stars look blue. Hence from the colour of a star we can, in principle, determine its temperature — although astronomers have more precise ways of determining the temperature os stars based on the presence of dark absorption lines in stellar spectra (Figure 17.3), the relative intensities of which vary with temperature.

At any temperature a hot body will emit radiation over a wide range of wavelengths and the overall visual effect, if there is one, depends on the relative amounts of the various parts of the spectrum. A body at the temperature of the Sun has its main visible emission in the green to red part of the spectrum — which makes it appear as a yellow star. A very hot star will have peak emission in the ultraviolet region; over the visible part of the spectrum there is high emission everywhere but more at the blue end than the red end — hence the appearance of bluish-white. A comparatively cool object, say at the temperature of boiling water, will emit infrared radiation that can be detected as heat but is invisible to the eye.

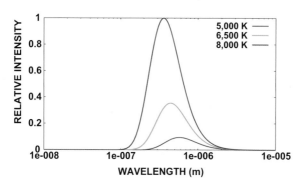

Figure 16.2 Relative radiation curves at 5,000 K, 6,500 K and 8,000 K.

The energy output as a function of wavelength for a body at a particular temperature was theoretically deduced in 1900 by the German physicist Planck, Max (1858–1947). In Figure 16.2 the Planck radiation curves, giving the relative intensity of radiation per unit area from a heated object for different wavelengths, are shown for the temperatures 5,000 K, 6,500 K and 8,000 K. First it will be seen that the peak of the curve moves towards shorter wavelengths for higher temperatures; the product of the peak wavelength and the absolute temperature is a constant — a relationship known as Wein's law. The other very noticeable feature is that the total energy emitted by the complete range of wavelengths, given by the area under the curve, increases very rapidly with temperature; theory shows that it is proportional to the fourth power of the temperature so it is more than 6.5 times greater at 8,000 K than it is at 5,000 K. The total amount of radiation coming from an object at a particular temperature is proportional to the area of the radiating surface so the luminosity of a star thus depends on the fourth power of its temperature multiplied by its surface area.

The brightness of a star as seen from Earth will depend both on it luminosity and its distance. A car headlight that is dazzling when seen close at hand seems much less bright when the car is at a great distance. There are various ways of estimating the distance of stars and if the brightness of a star is measured and its distance is known then its luminosity can be found.

16.2 What is a Young Stellar Object?

We have already used the term *protostar* to indicate an object that will eventually become a normal star. When a protostar is formed as an identifiable entity it is a large nebulous ball of cool gas and dust that slowly at first, and then ever faster, begins to collapse. Its initial radius will be somewhere in the range 1,000 to 2,000 au, or from 30 to 60 times the distance of Neptune from the Sun. At first the protostar is so transparent that the heat generated within it by its collapse can readily be radiated away so the temperature remains approximately constant, but as it becomes denser so it becomes more opaque and more of the heat within it is retained and it becomes hotter. Eventually, as the internal pressure builds up, it stops collapsing quickly and becomes a hot body at a temperature of a few thousand degrees. It is radiating energy and, as it does so it slowly collapses and its temperature increases. It may seem odd that a radiating body becomes hotter but the extra energy required, both to heat it up and to provide the radiation, comes from its collapse. In the usual terminology of astronomers a newly-formed star in this state is known as a *Young Stellar Object* (YSO).

When we refer to the temperature of a star we usually mean the temperature of the surface that we can see, so that the temperature of the Sun is about 5,800 K. However, the temperature is much higher inside the star and in the case of the Sun the maximum temperature within it is about 15 million K. So it is for a YSO; in its slowly evolving state it may have a surface temperature of, say, 4,000 K but it may be several million K internally and it will steadily rise as the surface temperature rises. Eventually the internal temperature reaches a critical value at which nuclear reactions can occur, involving the transformation of hydrogen into helium. This source of energy now provides what is radiated away, the slow collapse ceases, and the star is said to be on the *main sequence*. Thereafter its temperature and luminosity will vary comparatively little over a considerable period of time. The Sun is a main sequence star; it has been on the main sequence for about 4,600 million years and will remain on it for another 5,000 million years.

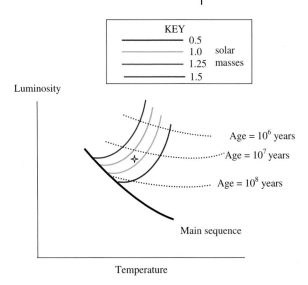

Figure 16.3 Pathways towards the main sequence.

Because the temperature and luminosity change little for a star on the main sequence it is impossible precisely to estimate its age. We can do so for the Sun by assuming that it has the age of the Solar System that can be dated in various ways not involving the Sun itself. But, when the YSO is on the slowly evolving path *towards* the main sequence both its temperature and luminosity *are* changing and can give an estimate of its age. In Figure 16.3 there is shown a schematic representation of the temperature and luminosity curves for YSOs of various masses in their final journeys to the main sequence as found from theory. If the temperature and luminosity of a YSO is measured then it is possible to determine both its mass and its age. For the observation represented by the cross the new star has a mass about 1.1 solar masses and an age of approximately thirty million years.

16.3 Detecting Disks

A YSO will be recognized by its temperature and luminosity but what can also be measured is the distribution of energy with wavelength, as shown in Figure 16.2, which is a well recognized curve for any particular temperature. In the mid 1980s it was found that for many YSOs the curve had an extra bump in the infrared part of the spectrum, which is outside the visible range but can be detected by instruments. We have already seen that low temperature sources, which emit infrared radiation, are very inefficient radiators in terms of rate of energy output per unit area but if they have a large area then they can give a considerable output. The bumps on the curves of new stars are called an *infrared excess* and they indicate the presence of a low-temperature, large-area radiating body accompanying the star. This is, quite reasonably, interpreted as being due to a large dusty gaseous disk encompassing the star.

From observations of many young stars it has been determined that the disks have a finite lifetime as there are very few disks for stars more than about three million years old. Lifetimes of from one to ten million years are inferred, with most being at the lower end of the range. The probable mechanism for dispersing the disks is the effect of the high emissions of both radiation and a stellar wind during what is termed the *T-Tauri* stage of the development of a YSO; the stellar wind at this stage can be a million times more powerful than the present solar wind. Other kinds of observations suggest what the disks are like. They seem to have masses mostly in the range one-hundredth to one-tenth of a solar mass with radii anything from tens to hundreds of astronomical units.

A widely-accepted theory of planetary formation involves the presence of disks so confirmation of their existence and the determination of their properties are of great interest.

Chapter 17

Planets Around Other Stars

Observe how system into system runs, what other planets circle other Suns

Alexander Pope (1688–1744)

17.1 Stars in Orbit

It is customary to refer to a planet being in orbit around the Sun, which is a correct statement in itself, but the mental picture that this evokes is of a planet moving around a *stationary* Sun. The Solar System is complicated, with many planets so, to simplify the discussion, we shall consider a star with a single planet. How should we describe the actual motion for this simple system?

What is really happening is that the two bodies, the star and the planet, are *both* in orbit. Their motion is such that they both go round a fixed point, the *centre of mass*, which lies on the line connecting them (Figure 17.1). The distances of each from the centre of mass is proportional to the mass of the *other* body; this means that if the star is 500 times as massive as the planet then the planet is 500 times more distant from the centre of mass.

The stellar orbit as seen in the figure is greatly exaggerated in scale for clarity of presentation. If the Solar System consisted just of the Sun and Jupiter then, since the Sun is 1,000 times as massive as Jupiter, the orbit of the Sun would be 1,000 times smaller, and the orbital speed of the Sun 1,000 times smaller, than that of Jupiter. The speed of Jupiter in its orbit is 12 km s^{-1}, which means that the speed of the Sun in its orbit would be 12 m s^{-1}. A similar situation would occur for any star with an accompanying planet.

These relationships can be expressed as

$$\frac{r_S}{r_p} = \frac{\nu_S}{\nu_P} = \frac{M_P}{M_S} \qquad (17.1)$$

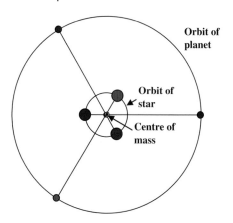

Figure 17.1 The orbits of a star and its planet.

where r, v and M indicate distance from the centre of mass, orbital speed and mass respectively and subscripts S and P indicate the star and the planet.

It is difficult, although as you will see later not impossible, visually to detect a planet in the immediate vicinity of a star since the light of the star would overwhelm the light reflected off the planet. However, if we imagine that the line of sight from the Earth was in the plane of the star's orbit then, as the star orbited, it would be seen to be first retreating from and then approaching the Earth in a periodic fashion. If this motion could be detected and measured then, although we could not see the planet directly, we could see a consequence of its presence and thereby be able to learn something about it.

17.2 Finding the Speed of a Star

Imagine being present at a Formula I Grand Prix race observing from the middle of a long straight. The cars approach with their finely-tuned engines screaming at a high frequency corresponding to the speed with which the engines are revving. They then draw level with the observing position and thereafter they are retreating. But — what has happened? The note from the cars has noticeably changed with the frequency now much lower. Have the drivers all throttled back? No, they have not, and what has been heard is due to a well know phenomenon in physics — the Doppler Effect.

Sound is a wave motion consisting of variations of air pressure propagating with a particular wavelength and a particular pitch (frequency). The Doppler Effect is the process by which an approaching source has a higher perceived pitch and a retreating source a lower one. Now the frequency, ν, and wavelength, λ, of a sound wave are related by

$$\lambda\nu = V \tag{17.2}$$

where V, the speed of sound, is the same for all frequencies. Hence, in terms of wavelength, the wavelength of an approaching sound source is shortened (higher frequency) and of a retreating source is lengthened (lower frequency). In a conceptual way we may consider that the approach of a sound source compresses the sound waves and makes them shorter while the retreat of the source stretches them out as illustrated in Figure 17.2.

What is true for sound is also true for light, which is another type of energy propagated in the form of waves. The light from a source that is moving away from an observer is seen with a longer wavelength, i.e. one towards the red end of the spectrum, than if it is at rest. Conversely, if it is moving towards the observer then the light appears to be of shorter wavelength i.e, towards the blue end. This then provides us with a way of detecting the motion of a star since a star is, after all, a source of light. However, we should not be able to

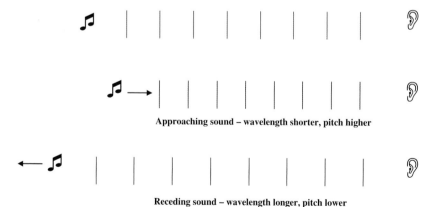

Approaching sound – wavelength shorter, pitch higher

Receding sound – wavelength longer, pitch lower

Figure 17.2 An illustration of the Doppler Effect.

Figure 17.3 Absorption (Fraunhofer) lines in the solar spectrum.

detect the motion just by looking at the overall colour of the light since the change would be far too small to detect in that way. Fortunately the spectrum of a star, just like that for the Sun, contains a large number of dark lines (Figure 17.3) corresponding to light that has been subtracted from the spectrum due to absorption by various kinds of atom as the light passed through the outer layers of the star. These act like markers for particular wavelengths of light and we can detect motion by the changes of wavelength of these lines.

To determine the speed of the star, v_S, either towards or away from the Earth requires a measurement of the changes in the wavelengths of the lines. We can then use the Doppler-effect formula

$$\frac{v_S}{c} = \frac{\Delta\lambda}{\lambda} \tag{17.3}$$

in which v_S is the component of the speed of the star along the line of sight, c, is the speed of light, λ the wavelength of a spectral line and $\Delta\lambda$ the change in the perceived wavelength. The speed of light and the wavelength of the line are known so that a measurement of the change of wavelength gives the speed of the star along the line of sight. If the change of wavelength is positive then the star is moving away from the Earth or, if negative, towards the Earth. During the period that stellar motions have been measured the measuring technology, using precise optical instruments called echelle spectrometers, has greatly improved and speed measurements now have a precision better than 1 m s^{-1}.

Of course there are a few complications in these measurements since the Earth itself is moving round the Sun and the centre of mass of the star-planet system may also be moving relative to the Sun. Nevertheless, it is possible to allow for these factors and explicitly to determine the speed of the star in its orbit.

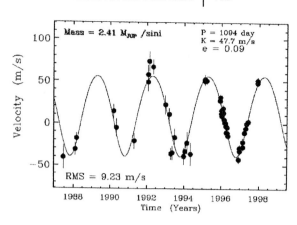

Figure 17.4 Stellar velocity measurements for 47 Uma.

Figure 17.4 shows some of the first observations of variable stellar velocities taken over a ten-year period for the star 47 Uma. The difference of the maximum and minimum of the curve fitted to the observation points gives twice the speed of the star in its orbit. The period of the fluctuation, 1,094 days gives the period of the stellar, and hence also of the planetary, orbit.

17.3 Finding Out About the Planet

Knowledge of the speed of the star in its orbit and its orbital period is not sufficient information to enable an estimate to be made of the mass and orbit of the planet. Another necessary piece of information is the mass of the star and, fortunately, we have a way of estimating this. In the Chapter 16 we introduced the idea of a main sequence star, one that is converting hydrogen to helium and which is in a long-lasting state. For main sequence stars the temperature, mass and radius are related so that if we know the temperature of a main-sequence star then we can also estimate its mass and its radius. The stars that have been found to possess planets are all main sequence stars so by measuring their temperatures their masses can also be estimated. As previously mentioned in Section 16.1, astronomers have a subtle way of determining the temperature of a star by looking at the relative strength of the absorption lines in their spectra corresponding to different atoms, but describing the process in terms of temperature illustrates the general principle.

From the theory of planetary motions developed by Newton, if the mass of the star, M_S, and the period, P, of the orbit are known then so is the semi-major axis, a, of the orbit and the speed of the planet, v_P, in that orbit. This comes from the relationships

$$P = 2\pi \sqrt{\frac{a^3}{GM_S}}, \qquad (17.4)$$

which gives a and then, assuming a circular orbit, dividing the circumference of the orbit by the period gives the speed

$$v_P = \frac{2\pi a}{P}. \qquad (17.5)$$

Equation (17.4) assumed that the star is much more massive than the planet and we are also assuming a circular orbit. With v_P, v_S and M_S known then from (17.1) the planet's mass, M_P, can be found.

All the analysis above assumes that the plane of the star-planet orbit is in the line of sight, which in general will not be true. The Doppler-shift measurements give the speed along the line of sight that will normally be smaller that the actual speed of the star. The relationship between the speed along the line of sight and the actual speed of the star is shown in Figure 17.5.

Since the estimate of the mass of the planet is proportional to the speed of the star if we underestimate the speed of the star then we also underestimate the mass of the planet. For this reason we are only able to estimate the *minimum planetary mass* and the true mass of the planet will be greater by an unknown factor.

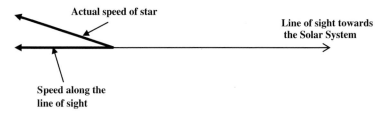

Figure 17.5 The speed along the line of sight is smaller than the speed of the star.

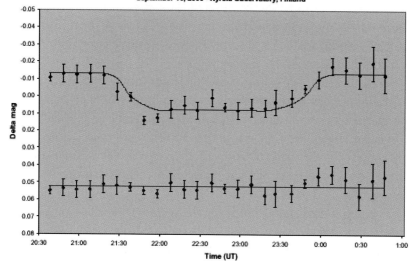

Figure 17.6 The light curve from the transit of an exoplanet over the star HD 209458. The lower line is from a check star. This observation was made by a group of amateur astronomers in Finland at the Nyrölä observatory.

There is a situation, by its nature rare, where an actual planetary mass can be estimated. This is when the line of sight is so close to the plane of the orbit that the planet actually transits the star — that is to say that the disk of the planet moves across the disk of the star. That this has happened is indicated by a diminution of the light as the planet makes its passage. An excellent example of such an observation is shown in Figure 17.6. The proportion of reduction in the brightness of the star gives the ratio of the cross sectional area of the planet to that of the star. Since the radii of main sequence stars are known from their temperatures it is possible in this way also to estimate the radius of the planet.

In March 2009 NASA launched its Kepler mission, a telescope that monitors about 145,000 stars in a fixed portion of the Milky Way galaxy, taking continuous photometer measurements to detect any transits that occur. The main objective of the mission is to detect planets of Earth-size, more-or-less, and to find how many of them are in a habitable zone where biological systems might be present. By early 2013 they had discovered 2,712 possible candidates. NASA scientists have claimed on the basis of their present observations that in the Milky Way galaxy there may be billions of near Earth-size planets — some smaller and some bigger — that may orbit in regions that could support life.

In the description so far of the determination of planetary orbits and of minimum planetary masses the assumption has been made that the orbits are circular rather than being general ellipses. If the orbit is an ellipse then the curve fitted to the observed stellar speeds is a somewhat distorted version of that previously shown and the degree of distortion enables the eccentricity of the orbit to be found (Figure 17.7).

The number of stars found to possess one or more accompanying planets, known as *exoplanets*, increases steadily with time and is likely to do so for the foreseeable future. Not all stars have planets or, more precisely, detectable planets. The easiest planets to detect by the Doppler-effect method are those with large masses in close orbits. A large planetary mass makes the motion of the star greater and hence easier to detect. Close orbits give shorter orbital periods and higher orbital speeds thus enabling a complete cycle of the planetary motion to be found more quickly and making the Doppler effect larger and hence easier to measure. Early measurements could not detect planets of less than about the mass of Saturn but instrumental

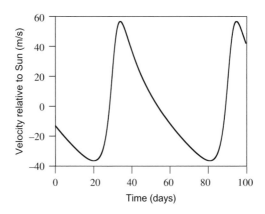

Figure 17.7 A typical velocity profile for a star with a planet in an eccentric orbit.

improvements have now led to the detection of planets with just a few times the mass of the Earth. For a star of, say, solar mass with a planetary orbit of radius 30 au the period would be more than 160 years, so it would take several tens of years just to detect the presence of the planet. For this reason we cannot estimate with any degree of certainly what proportion of stars has planetary companions. The estimates of the proportion have steadily increased with time and some workers in the field have given estimates of 30 percent or more.

17.4 Inferring the Presence of Planets from Dusty Disks

The discovery of dusty disks around young stars was described in Section 16.3. Actually, dusty disks around older stars are also very common — for example, the Sun, has a dusty disk. In parts of the world, such as deserts, where astronomical viewing conditions are ideal, the Sun's dusty disk can be seen with the unaided eye. Just after sunset or just before dawn, when the sky is dark but the Sun is not too far below the horizon, the disk is seen as a faint band of light in the sky called the *zodiacal light* (Figure 17.8). Dust in the Solar System is concentrated near the ecliptic and sunlight scattered from it gives the zodiacal light.

The fact that this dust exists gives information about processes happening in the Solar System. Due to a phenomenon known as the *Poynting–Robertson effect,* solar radiation causes small particles to spiral inwards towards the Sun. Thus a grain of sand orbiting the Sun at 1 au would be absorbed by the Sun in about one million years. This shows that the dust must be constantly replenished from some source, or sources, otherwise during the lifetime of the Solar System it would have disappeared. In the case of the Solar System we know that the sources are the dust being shed from comets and small debris from the occasional collision of asteroids

This dust emits infrared radiation over wavelengths appropriate to its temperature but the Sun's disk could not be detected by an observer outside the Solar System because the solar dust disk is so unsubstantial. Because fine dust presents a large surface area even for a small total masses of material, a dust disk with one Earth mass

Figure 17.8   The zodiacal light (NASA/courtesy of nasaimages.org).

would be easy to detect; the actual mass of the Sun's disk is a tiny fraction of an Earth mass.

There are many stars, not young but much younger than the Sun, which possess easily detectable dusty disks and it follows that these stars must possess accompanying bodies that are rich sources of dust. There have been detailed infrared studies of a number of such stars by a group based at the University of St Andrews in Scotland. They found that the star Fomalhaut, 25 light years from the Sun with mass 2.1 $M_\odot$[1], has a dusty disk that is cleared out in its middle region, suggesting the presence of planets that are sweeping up the dust. There is also bright ring at about 40 au from the star that may be a region rich in comet-like bodies.

Many other stars with dusty disks have features suggesting the presence of particular types of body. For example, Vega, also with mass 2.1 $M_\odot$ and distant 25 light years, has a cleared-out centre of

---

[1]The symbol $\odot$ represents the Sun so $M_\odot$ is the solar mass.

its disk and a bright concentrated source at about 80 au, which may be emission from a dust cloud surrounding a major planet. Such a planet would not be detectable by Doppler-effect methods because of its long 500 year orbital period.

17.5 Direct Imaging of Exoplanets

In 2008 the St Andrews astronomers reported their observations of a star HL Tau, which is surrounded by a dusty disk. The radio image of the star showed the presence of a planet with estimated mass 14 times that of Jupiter (Figure 17.9); a body of this mass would usually be described as a *brown dwarf*. There seems to be another smaller radio source shown in the figure, although this was not identified as a planet. A nearby star, XZ Tau — a binary or possibly three-star system — is thought to have passed by HL Tau about 1,600 years ago and it was suggested that this passage had 'triggered' the formation of the planetary companion.

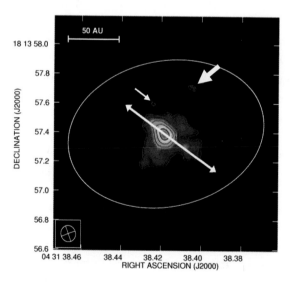

Figure 17.9 A 1.3 cm false-colour map centred on HL Tau. The large arrow marks the reported body. The smaller arrow may indicate the presence of a smaller body. The oval represents the limit of the gaseous disk surrounding the star. The arrows from the centre mark the directions of jets of hot gas.

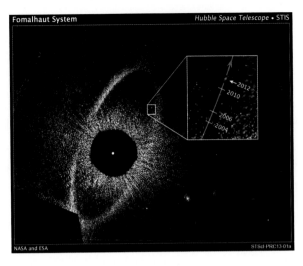

Figure 17.10   The dust disk around Fomalhaut. The enlarged image of the box shows the planet's position in 2004, 2006, 2010 and 2012 (NASA/ESA).

In 2008 the American astronomer, Paul Kalas, showed the first optical image of a planet orbiting the star Fomalhaut. In 2005 it was inferred from a sharp inner boundary of Fomalhaut's dust disk that a planet was present. Photographs of the dust disk taken in 2004, 2006, 2010 and 2012 show the movement of the planet over this period (Figure 17.10). Its mass is indeterminate and the best estimates of its orbital characteristics are that its semi-major axis is about 115 au, corresponding to a period of 872 years, and orbital eccentricity approximately 0.11 — but these are uncertain values.

At the same time as the Fomalhaut announcement the Gemini Observatory released an image showing three planets around HR8799, a star of mass 1.5 $M_\odot$ in the constellation Pegasus. These planets, of masses estimated as 10, 10 and 8 Jupiter masses are at distances 25, 40 and 70 au from the star, respectively.

There are other examples of direct imaging of planets. What they have in common is that the planets are at distances from their stars that are not amenable to Doppler-effect measurements, with some distances much further that Neptune's distance from the Sun. We shall see, this has an important implication for theories of planetary formation.

Table 17.1   Characteristics of a sample of exoplanets

Star	Minimum mass of planet (Jupiter units)	Period (days)	Semi-major axis (au)	Eccentricity
HD 187123	0.52	3.097	0.042	0.03
τ-Bootis	3.87	3.313	0.0462	0.018
51 Peg	0.47	4.229	0.05	0.0
υ-Andromedae	0.71	4.62	0.059	0.034
	2.11	241.2	0.83	0.18
	4.61	1266	2.50	0.41
HD 168443	5.04	57.9	0.277	0.54
16 CygB	1.5	804	1.70	0.67
47 Uma	2.41	3.0 years	2.10	0.096
14 Her	3.3	1619	2.5	0.354

17.6 The Characteristics of some Exoplanets

A sample of exoplanets that have been detected is given in Table 17.1. There are several points of interest in this table. The first is that there are a number of orbits that are very close to stars. The nearest planet to the Sun is Mercury at a distance of 0.4 au. The first four entries in the table give orbits between about one-sixth and one-tenth of the Mercury distance. A second point is that υ-Andromedae has a *family* of planets, at least the three that have been detected. When there is more that one planet, say three, the star velocity values have to be fitted to a complicated curve which is the sum of three simple curves with different periods and different amplitudes[2]. A final point of interest is that some planetary orbits have high eccentricity, much higher than those in the Solar System. The planetary orbit of highest eccentricity in the Solar System is that of Mercury, 0.206. The highest eccentricity in the table is 0.67 and there are some exoplanet orbits with even higher values.

All these observations of exoplanets are relevant to theories of the origin of the Solar System. It would be extraordinary if exoplanets

---

[2]The amplitude of the curve is one-half of the difference between the maximum and the minimum.

were formed in some way different to the way that solar-system planets were formed. If that is accepted then any viable theory for the origin of the Solar System must, perforce, also be able to explain exoplanets and their characteristics.

17.7 Exoplanets going Backwards

In Section 8.2 the requirements were listed for a successful theory of the origin of the Solar System, given the state of knowledge prior to 1960. Included in this list was an explanation of the $7°$ tilt of the solar spin axis. It presented a particular problem for any nebula theory, such as that of Laplace (Chapter 9), but not a serious one since if a star had passed close to the Solar System at some time then it might have perturbed the planets enough to shift the plane of their orbits by the required angle.

In 2009 a discovery was made that astonished the astronomical world — an exoplanet, WASP-17b, was found in a retrograde orbit, which is to say that the angle between the spin axis of the star and the rotation axis of the planet was greater than $90°$. The WASP (Wide Angle Search for Planets) programme is run by a consortium of UK universities with the aim of detecting transiting exoplanets. When they found the characteristic dip in intensity as the planet crossed the disk of the star (Figure 17.6) they also looked at the spectral composition of the light that, through what is known as the Rossiter-McLaughlin effect, could reveal the relationship between the stellar spin axis and orbital plane of the exoplanet. A simple basis for the Rossiter-McLaughlin effect is shown in Figure 17.11. Assuming for now that the star is at rest relative to the observer, with the spin direction shown the right-hand side of the star, moving towards the observer, is blue shifted and the right hand edge, moving away from the observer, is red shifted. The observer cannot see these fine details but does see that spectral lines are widened by this combination of red and blue shifts but there is no apparent change in the radial speed of the star. In Figure 17.11a, with a prograde orbit for the planet, part of the blue shifted light is first removed and the spectral line appears to have become red shifted giving an apparent

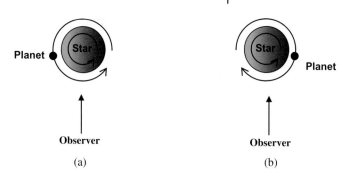

Figure 17.11 The Rossiter-McLaughlin effect for (a) a prograde orbit and (b) a retrograde orbit.

motion of the star away from the observer. Later there is removal of part of the red shifted light, giving an apparent blue shift of the spectral line suggesting that the star is approaching the observer. In Figure 17.11b, with a retrograde orbit for the planet, the apparent blue shift appears before the apparent red shift.

This brief description of the Rossiter-McLaughlin effect just gives the basic theory of the mechanism. The star does not have to be at rest relative to the observer and from careful analysis of the observations estimates can be made of the inclination of the exoplanet orbit relative to the spin axis of the star.

In 2010, at a National Astronomy Meeting of the Royal Astronomical Society held at the University of St Andrews, it was announced that of 27 exoplanets for which inclinations had been determined six (22%) had been found to have retrograde orbits. These new observations impose a serious constraint of theories of the origin of planetary systems in general and of the Solar System in particular.

Chapter 18

What a Theory Should Explain Now

It is a capital mistake to theorize before one has data.

Arthur Conan Doyle (1839–1930)
The Memoirs of Sherlock Holmes. The Crooked Man

18.1 The Beginning of the 21st Century

Although the hope was often expressed that exploration of the Solar System by spacecraft would lead to new understanding of how the system began, this has not happened. Naturally our knowledge of the Solar System is immeasurably greater than it was in 1960 but that knowledge has not helped greatly in the task of finding a theory of its origin. For example, it is of great interest to know that Jupiter's satellite Io has active volcanoes but it does not indicate how that body was formed. Probably the most helpful information we have had from space research is of a chemical nature. The chemical similarities and differences between lunar rocks and those on Earth might, perhaps, be saying something about the origin of the two bodies but might equally be indicative of events that happened after the bodies were formed. This is a constant problem in trying to interpret the results of space research. The Solar System has been in existence for four thousand five hundred million years and it has been an eventful period. We can see on Earth, and on other bodies, the results of bombardment with large projectiles but other cataclysmic events may have taken place, leaving clues that are difficult to decipher.

The astronomical observations made in the last ten years of the 20th century have been far more helpful. The observations of exoplanets enable us to rule out of contention any theory that would make the Solar System a unique or rare example of its kind. Events

165

of low probability — for example, as postulated by Jeans and Jeffreys — can be discounted on purely statistical grounds without considering all the detailed criticism of the mechanisms they proposed. The evidence from dusty disks around older stars (Section 17.4) and the direct imaging of some exoplanets (Section 17.5) indicate that any theory that could not produce planets at distances of a hundred, or more, au from the parent star should have a cloud of uncertainty hanging over it.

The Solar System has many complex features and it is expecting a great deal of any theory that it should be able to explain them all. We should distinguish between those features that are mandatory for any theory to explain and those that can be put aside as possible evolutionary features. There may be differences of opinion in the categorization of features. Nevertheless, there will now be given a selection of features that should be addressed by any plausible theory of the origin of the Solar System, given the data that is now available.

18.2 The Sun and its Properties

Although the search for possible theories is not a competition — at least it should not be for objective scientists — it still seems a little unfair that monistic and dualistic theories are not presented with the same challenges. Inevitably a monistic theory, that is producing the Sun and the planets from the same pool of material, will be required to give a slowly spinning Sun. Dualistic theories assume a pre-existing Sun and duck the issue of slow spin. There is the fact that slowly spinning solar-type stars are a matter of observation and a dualistic theorist will clearly use that argument. Nevertheless, to go some way to establishing a level playing field, any dualistic theory that is accompanied by some ideas about the Sun's origin and explains its slow spin should be given some brownie points. Equally, a monistic theory *must* convincingly explain the slow spin of the Sun.

The 7° tilt of the solar spin axis, which could also be interpreted as a 7° inclination of the planetary orbits, could conceivably be an evolutionary feature but extra brownie points should be given for an explanation of this property.

18.3 Planet Formation

This might seem a rather trite and obvious feature for a successful theory but achieving the formation of planets has proved to be quite a challenging theoretical exercise. Because of the lesson learned from the Jeans theory the planets must be made from cold material, where the term 'cold' is to be judged by stellar standards rather than by the standards of the kitchen.

The two kinds of planets, terrestrial and major, occupy different regions of the Solar System. It is possible that, for example, planets are formed as gas giants and later evolve into terrestrial planets, in which case division into the two types is an evolutionary feature. Some theories may naturally form both types of planet but for those giving only gas giants some explanation of terrestrial planets would obviously be desirable if not essential.

The model for planetary formation must be capable of explaining exoplanets as well as the planets of the Solar System so the mechanism must be robust, one that does not depend on fine tuning the parameters of the model — temperature, density etc. It must be capable of giving very close orbits, much closer to stars than Mercury is to the Sun, and also very distant orbits, much greater than Neptune's distance from the Sun. Finally, although many exoplanet orbits are close to circular there are some with high eccentricities and some explanation for this characteristic is also required.

For the Solar System the orbits of the planets are direct and nearly, but not quite, coplanar. The fact that a significant proportion of exoplanets have retrograde orbits is clearly something that any theory *must* explain. If it is not *directly* a feature of the proposed process by which the planets form then mechanisms must be found that would transform some originally-prograde orbits into ones that are retrograde. A feature that is still not determined, but which would be of critical importance, is whether or not for systems of exoplanets they would always be orbiting in the same sense — all prograde or all retrograde.

18.4 Satellite Formation

There is no evidence that exoplanets possess satellites so it is possible that satellite formation may not be a essential requirement of a general theory. However, regular satellites occur for *all* the major planets of the Solar System so it seems much more likely that satellites are a usual, perhaps even inevitable, concomitant feature of planet formation, so some model of their formation is desirable.

18.5 Asteroids, Comets and Dwarf Planets

Some theories have asteroids and comets as precursors of planet formation. For such theories it is therefore required that the origin of these basic bodies should be explained. For other theories, planets form directly from diffuse material without the intermediate stage of producing smaller solid bodies. Again, in the interest of having parity of treatment for two different approaches, the latter theories should be considered more favourably if, in addition to giving a plausible model for planet formation, some process for formation of the smaller bodies is also given.

There seem to be two distinct sources of comets — bodies within the Kuiper Belt that are perturbed by Neptune to give short-period comets and those in the Oort Cloud that are perturbed by bodies external to the Solar System. An explanation of these two sources of volatile-rich bodies is clearly desirable.

The dwarf planets all have masses and radii that correspond to those of smaller satellites of the major planets. The question of their origin, and whether they were produced by the same, or a different, mechanism that produced satellites is clearly a matter of interest.

18.6 Concluding Remarks

There has been such a variety of theories proposed over the years, with different assumptions built into them, that the facts that their authors try to explain are also very variable. A prominent American science historian, Stephen Brush, concludes, perhaps cynically, that every individual's list of facts to be explained consists of just that set of features that his or her theory can explain. This, after all, is the basis of the quotation from Karl Marx that heads Chapter 8. There may be an element of truth in Brush's comment but everyone would agree that the more that can be explained the better and that there are some features of the Solar System that are so basic, e.g. planet formation, that they must be on everyone's list.

Some of the features mentioned in this chapter should be regarded as mandatory for any plausible theory — for example producing planets and for them to be at a wide range of distances from their star, with the possibility that some of them will be in retrograde orbits. Other features may be regarded as being less basic, although any theory that can explain them will earn and deserve greater credibility.

Part VI

The Return of the Nebula

Chapter 19

The New Solar Nebula Theory: The Angular Momentum Problem

I shall return.

General Douglas MacArthur (1880–1964)

19.1 A Message from Meteorites

Studies of meteorites in the 1960s indicated that many of their characteristics could be interpreted in terms of their origin as condensations from a hot vapour. We are familiar with pictures of molten rocks pouring out of a volcano. Now we have to think of even higher temperatures where the rocky, mainly silicate, material does not just melt but turns into a vapour — as water turns into steam when it boils. When the silicate vapour cooled it first liquefied and when this happened it produced small spherical droplets, just as a fog produced by condensing water vapour consists of tiny spherical droplets of liquid water. In both cases, for the liquid silicate and the water, the agency causing the spheres to form is surface tension, the energy associated with the surfaces of liquids. All physical systems tend to adopt a state of lowest energy and for a given volume of liquid a sphere gives the least surface area, and hence the lowest surface energy. This only occurs for tiny droplets of liquid; for larger volumes of liquid, energy due to the local gravitational field dominates and the liquid will adopt a configuration of lowest total energy — gravitational plus surface. Having a silicate vapour explains the formation of the chondrules — solidified droplets — found in chondritic meteorites, which later become embedded in fragments of other minerals and eventually were compressed into the form of a rock

There is an important difference between water vapour and a silicate vapour. The description of water as H_2O is familiar — one atom of oxygen linked to two atoms of hydrogen. Water vapour consists of large numbers of H_2O units all moving around freely in space and when they condense to form either water or ice they do so with no change of chemical identity. By contrast, when a rock is vaporized it breaks up into units that are smaller than the chemical units that constituted the minerals in the rock. For example, the mineral forsterite, a form of olivine that is an important major component of the Earth's mantle, has the chemical formula Mg_2SiO_4 — a combination of two atoms of magnesium, one of silicon and four of oxygen. However, there are no identifiable discrete units in the solid mineral, separated from other units, corresponding to this formula. Rather the atoms are arranged in a three-dimensional framework and the chemical formula describes the contents of a basic unit of the framework that, when packed together, creates the whole mineral. Vaporizing the material breaks up the framework to produce small very stable entities that then become the components of the vapour. Thus forsterite will break up as

$$Mg_2SiO_4 \rightarrow 2MgO + SiO_2$$

in which, chemically, the right-hand side is two molecules of magnesium oxide and one molecule of silicon dioxide (silica). When the rock vapour is cooled the individual units reconnect to form minerals that may be different from the original minerals that produced the vapour in the first place. Just what new minerals are produced depends on the rate at which the rock vapour is cooled and also on how diffuse or concentrated the vapour is. In general, as the rock vapour cools so the first minerals to condense out are those that boil at the highest temperatures followed by those that boil at progressively lower temperatures. Many of the characteristics of meteorite minerals are explicable in terms of 'condensation sequences' from a hot vapour. Another clue was that small crystals within cavities in meteorites showed all the characteristics of having been deposited directly from the vapour phase.

The message was clear — at some time in the history of the Solar System the material from which at least some meteorites formed had been in the form of a vapour. This stimulated the thought that this vapour could have been a nebula of the type first envisaged by Laplace. It was realised that Laplace's model had foundered on the angular momentum problem, mainly manifested in the slow spin of the Sun. However, it was argued, more was known in the twentieth century than in the eighteenth and theorists were confident that somehow or other the problem could be solved. The nebula had returned and the development of the new Solar Nebula Theory (SNT) had begun.

19.2 Mechanical Slowing Down of the Sun's Spin

The first burning problem to be tackled was just how to transfer angular momentum from the core of the collapsing nebula, which would eventually become the Sun, to the disk of material left behind within which the planets would form. The first idea in this direction came in 1974 from the Cambridge astronomers, Donald Lynden-Bell and Jim Pringle. There are various reasons why the nebula would not collapse in a well-ordered, quiet and smooth way. The very process of collapse could introduce some turbulent motion into the nebula. Again, if the core became hot then the disk could be stirred up by being heated just as water in a heated saucepan becomes agitated. Another possibility is that material from outside falls onto the disk, creating disturbances in the way that a pond is disturbed by a stone thrown into it. In summary, the collapsing cloud is likely to become turbulent, a condition that is described more fully in Chapter 27. In a material with turbulence, mechanical energy turns into heat and this heat will then be radiated out of the system and lost. The net effect is that in a turbulent medium the energy of motion steadily reduces.

In Chapter 9, where the idea of angular momentum was first introduced, there was given the important principle that in an isolated system angular momentum remains constant. So it is for the collapsing nebula — the angular momentum must remain constant but, because of turbulence, the energy of motion must reduce. How does the material rearrange itself to satisfy these two conditions? This was the problem solved by Lynden-Bell and Pringle. They

Figure 19.1 Edge view of the nebula showing the motion of material due to turbulence.

showed that material on the inside of the disk must move further inwards while material on the outside must move further outward (Figure 19.1). This means that inner material loses angular momentum while outer material gains it, which is tantamount to an outward transfer of angular momentum from the inner collapsing core to the disk — just what is required.

This process is helpful but it does not solve the problem completely. The way that the inner material moves is by gradually spiralling inwards. At all times it is orbiting at a speed corresponding to circular motion around the inner mass. When it joins the central mass it would be in free orbit at the equator just as though it was an orbiting planet. A body formed in this way would nearly, but not quite, be at the point of spinning so fast that it would fly apart. That is nothing like the state of the present Sun so there must be some other mechanism available to transfer more angular momentum from the core to the disk.

19.3 Magnetism Gives a Helping Hand

The majority of large astronomical bodies, from the size of planets upwards, have associated magnetic fields. That of the Earth shows itself by its action on a compass needle that always points northwards[1], which has been a navigational aid since the time of ancient China.

Figure 19.2 shows a simplified picture of the Earth's magnetic field with the lines, called *flux lines*, giving the direction of a

---

[1] A compass needle does not point towards the geographical north pole but towards the *magnetic north pole* that moves around but is now situated in the far north of Canada.

Figure 19.2   Magnetic flux lines for the Earth.

compass needle at various points in the vicinity of the Earth. Of course the flux lines are not physical realities — they are a conceptual construct that enables us to model the behaviour and action of a magnetic field. The north-poles of the little bar magnets in the figure are white.

The planet Jupiter has a larger magnetic field than that of the Earth and the Sun's magnetic field is larger than that of Jupiter. The magnetic fields associated with newly-formed stars can be even greater than that of the Sun and the model for the outward transfer of angular momentum, now to be described, considers a young rapidly-spinning star with a magnetic field, accompanied by a disk. A mechanism for transferring angular momentum by a magnetic field was suggested by the British astrophysicist Fred Hoyle (1915 — 2001) as early as 1960. His idea depended on having the central collapsing core and the edge of the disk at a temperature high enough for the material to be ionized. The ionization process means that atoms are broken up into negatively charged electrons and the rest of the atoms (called *ions*) with a positive charge. An intimate mixture of electrons and positively charged ions is called a *plasma* and,

Figure 19.3   Fred Hoyle.

because they consist of charged particles, plasmas are good conductors of electricity.

If a magnetic field exists within highly conducting plasma then it will be *frozen in*, meaning that if the plasma moves then the flux lines will move with it. Hoyle envisaged the central collapsing core of a nebula separated by a gap from the inner edge of the surrounding disk, also consisting of ionized material, with magnetic flux lines linking them (Figure 19.4). The flux lines behave like rubber bands and if they are stretched they try to shorten themselves, since this reduces the energy associated with the magnetic field. The flux lines shown in the figure are frozen into both the core and the disk meaning that they are firmly attached at each end. The core spins more rapidly than the disk so it stretches the flux lines. To shorten themselves the flux lines pull inwards at each end and from the figure it is clear that this will have the effect of slowing down the spin of the core and speeding up the spin of the disk — effectively transferring angular momentum outwards.

The problem with the Hoyle model is that the magnetic field that he assumed was too small so that the field lines became greatly stretched before they could exert a significant force on the core and disk. It is as though the rubber bands were very thin and easily stretched. Theory shows that, before the flux lines stretched to the

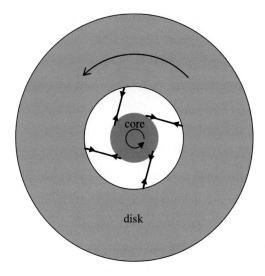

Figure 19.4   Flux lines linking a collapsing core to a disk.

extent required, the energy of the magnetic field would be reduced by rupture of the flux lines between the core and disk and then reconnection to give new shorter flux lines. The flux lines would be continuously stretching and breaking and there would never be sufficient force acting on the core and disk to transfer angular momentum.

19.4 A Modification of the Hoyle Mechanism

There have been a number of newer ideas for transferring angular momentum outwards by the use of a magnetic field that differ in the initial assumptions they make. Here we just describe one that has been developed in some detail by P.J. Armitage and C.J. Clarke in 1996.

The core is taken to be in the form of a very active pre-main-sequence star with a very high magnetic field, perhaps one thousand times that of the Sun. The disk is in orbit around the central mass with material in the inner part of the disk orbiting at a rate faster than that of the core. The disk material further out is orbiting at a lower rate, just as outer planets orbit the Sun more slowly than the

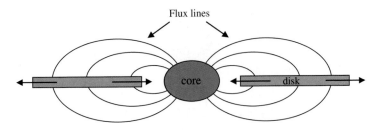

Figure 19.5 Inner disk material moves inwards to join the core while outer disk material moves outwards.

inner ones. In Hoyle's original model the magnetic linkage was just to the inner region of the disk but now magnetic linkage is assumed to occur between the core and all parts of the disk, as illustrated in Figure 19.5 in edge view. The magnetic flux lines linking the core to the more slowly rotating outer material transfer angular momentum outwards just as Hoyle had described. Consequently that material moves outwards. Because the magnetic field is so strong the flux lines behave like very strong rubber bands that have to be stretched very little to give the required forces and therefore do not break and reconnect. The more rapidly orbiting material in the inner part of the disk is also connected to the core by flux lines but these give a different form of behaviour. The inner part of the disk is now assumed to become rather turbulent and material from this part of the disk flows along the flux lines to join the central core.

This model of transfer of angular momentum through the action of a magnetic field requires several conditions to be satisfied. The core must generate a field of sufficient strength and, although strong surface fields have been detected on some young stars, it is not completely certain that the field would be of a form that would enable a strong linkage to occur. Another requirement is that the disk material should be hot enough to be conducting at some considerable distance from the star. That should not be a problem at the inner edge of the disk, which is close to the star and fully exposed to its radiation. Finally the disk would have to be reasonably quiet and not too turbulent as otherwise it may not be possible to set up the required pattern of flux linkage. Another problem with the particular model described here is that the initial state taken for the core was far from being at the point of rotational disruption, which we have already seen is what happens through the action of the Lynden-Bell and Pringle mechanism. The important process of attaining this initial state has still to be described.

19.5 Slowing the Sun's Spin

Even if the transfer of angular momentum outwards can be achieved by some means or other then it is not very likely that the transfer process would leave the Sun spinning at its present very slow rate — with a period of about 27 days. We now consider another mechanism depending on a solar magnetic field that would slow down the spin of the Sun without reference to any external body such as a disk. This relies on the interaction between the solar magnetic field and the *solar wind*, a stream of energetic particles — mostly electrons and protons — leaving the Sun at a speed of about 500 km s^{-1}. The rate of loss of mass of the Sun due to the solar wind is two million tonnes per second; to put it in perspective this rate of loss would cause the Sun to lose one ten-thousandth of its mass over its 10,000 million year main-sequence lifetime.

When streams of charged particles move in the presence of a strong magnetic field they can become coupled together so that the charged particles move along a helical path with a flux line as an axis of the helix (Figure 19.6).

When the Sun, or any star, spins, the magnetic field also spins with it like a rigid structure, and this would apply to all the imaginary flux lines that define the field. This means that the solar-wind particles would move outwards along a flux line at a *constant angular speed*. How far the particle moves along the flux line before it breaks free depends on the energy of the particle and the strength of the magnetic field. The more energetic the particles the sooner they

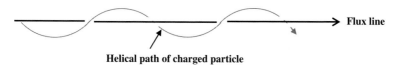

Helical path of charged particle

Figure 19.6   The helical path of a charged particle coupled to a flux line.

break free; the stronger the field the later they break free. The consequence of this process is clear from Equation (9.1). The distance from the star (r) increases but the angular speed (ω) is unchanged, which means that the angular momentum associated with the escaping particles is increasing. However, the total angular momentum of the system has to be conserved so there is a compensating decrease in the angular momentum of the remaining bulk of the Sun — i.e. it spins more slowly.

With the present rate of loss of matter due to the solar wind and the present strength of the Sun's magnetic field the effect is a weak one and would not have removed much angular momentum over the lifetime of the Sun. However, it is likely that the early Sun was much more active than it is now and, for a period of a million years or so, the rate of mass loss could have been more than one million times the present rate and the field one hundred times greater than at present. It is possible that the Sun retains just a few percent of its original angular momentum.

19.6 The Stellar Angular-Momentum Conundrum

It is not clear how a compact object, destined to form a star, can come about by a process in which material initially spirals inwards to join a central core. A condensing body produced in this way would always be on the point of disruption and it must lose angular momentum to collapse further. However, the high magnetic fields and stellar winds that could achieve this loss of angular momentum, and subsequent collapse, can only occur if it is a hot compact body — which can only come about if it collapses.

Work continues on this topic of outwardly transferring angular momentum from core to disk. It is believed by those working on the SNT that, while there may still be some further work to be done in this area, the essential problem of angular momentum transfer can be solved by the action of magnetic fields and that the essential starting point of a relatively slowly spinning core and a surrounding extensive disk, containing most of the angular momentum of the system can be achieved. If that is true then the next problem to be considered is that of forming planets in the disk.

Chapter 20

Making Planets Top-Down

A great while ago the world began.

William Shakespeare (1564–1616) *Twelfth Night*

20.1 A Massive Disk

While the tidal model put forward by Jeans did not stand the test of time, the theory that underpinned the model is completely valid. For example, a gaseous filament would break up into a string of blobs and any blob that had greater than the Jeans critical mass would collapse to form a condensed body; the fact that this happens will be demonstrated by a simulation in Chapter 29. There was nothing wrong with Jeans' theoretical work — the failure of his model was due to the scenario in which he applied it.

The kind of behaviour that Jeans found for a filament would also occur in a spread-out thick disk of material. It too would be gravitationally unstable and break up into a number of blobs, in this case distributed over the disk rather than strung out in a line (Figure 20.1). If these blobs had greater than the Jeans critical mass then they could form condensed bodies.

The first idea about making planets with the SNT was through such a process. The disk was taken as having sufficient density and areal density (mass per unit area) to produce blobs that would be above the Jeans critical mass and so condense to form major planets. Equation (11.2) shows that the critical mass depends on both the density and temperature of the material. The higher the density the *lower* is the critical mass, the higher the temperature the *higher* is the critical mass. In order to produce critical masses equal to that of, say, Jupiter with material at a temperature that could vaporize silicates required absurdly high densities — so high that the mass of

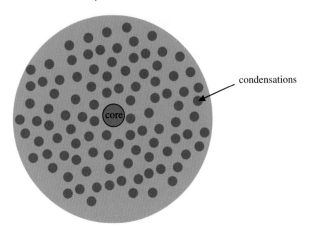

condensations

core

Figure 20.1   The spontaneous break-up of a disk into protoplanets.

the disk would have been very much greater than the mass of the central core that was to form the star. In 1978 the Canadian astronomer, A.G.W. Cameron, who was investigating this process for producing planets, wrote — "At no time, anywhere in the solar nebula, anywhere outwards from the orbit of formation of Mercury, is the temperature in the unperturbed solar nebula ever high enough to evaporate completely the solid materials contained in interstellar grains. For some time a number of people have argued that the entire solar nebula started out at a high temperature and cooled while solids underwent a sequence of condensation processes. In fact, there is no available energy source for any such high temperatures to have been initially present." The idea of a very hot nebula had to be abandoned and consequently the whole *raison d'être* for the resurgence of nebula ideas was lost. However, by this time the SNT had become firmly established as the most plausible model for forming the Solar System and work on it continued unabated. Other ways would have to be found for producing the vapours that meteorite observations required — but that was not an immediate concern of the solar-nebula theorists who were more engrossed in the problem of producing planets.

This way of producing major planets by direct condensation of the disk is known as 'top-down'. The idea behind the name was that there is a hierarchy of sizes of objects in the Solar System — dust, small asteroids, large asteroids, satellites, terrestrial planets and major planets, with the major planets at the top of the list. Top-down formation implies that the largest bodies, the major planets, are first produced and the formation of other smaller bodies is through their subsequent disruption.

20.2 The Problems of Top-Down Processes

Even with a low-temperature disk, say in the range 10–100 K, the density required to produce major planets was still so high that the disk had to have a mass that was similar to that of the core. The fragmentation of a gravitationally unstable disk would have produced not a small number of major planets but a very large number of such bodies. If the conditions were right in one part of the disk to produce a major planet then they would also be right in many other parts of the disk. This raised the question of the disposal of superfluous material, since a great deal of energy would be required to expel almost the mass of the Sun, from the Solar System. No possible source for such an amount of energy was available.

Later, when disks were detected around young stars, the observations did not indicate the existence of very massive disks; the maximum mass of a disk is thought to be about 0.1 $M_\odot$. Support for the top-down approach declined and other ways of producing planets from a disk were considered.

Chapter 21

A Bottom-Up Alternative

Tall oaks from little acorns grow.

David Everett (1769–1813)

21.1 A Summary of the Bottom-Up Approach

The antithesis of the top-down process is to start the process of forming planets, and other bodies, with the dust contained within the disk. This would have constituted about one percent of the mass of the disk and it would have been in the form of very tiny grains. The proposed stages of planet formation in this way are as follows:

(1) The dust gradually settles down under gravitational forces into the mean plane of the disk to form a thin dust carpet.
(2) Just as for the gas disk described in Chapter 20, the dust carpet is gravitationally unstable and breaks up into a large number of small compact solid bodies. Following the notation of Chamberlin and Moulton (Section 10.2) these bodies are called *planetesimals*.
(3) The planetesimals aggregate to form larger solid bodies. In the inner part of the Solar System they form the terrestrial planets. In the outer part of the system they are the cores of the major planets.
(4) The major planets acquire gaseous envelopes that, at least for Jupiter and Saturn, account for most of their masses.

Each of these stages will now be described.

21.2 Forming a Dusty Carpet

The endpoint of this stage of the bottom-up process is to produce a solid layer in the mean plane of the disk, as shown in Figure 21.1,

Figure 21.1 The gravitational settling of dust to form a dust carpet.

consisting of the dust that has settled through the effects of gravity. The dust particles in the nebula disk would have been very tiny, with dimensions typically about one micron. They would have been constantly buffeted in all directions by the molecules of the surrounding gas which exerted on them much stronger forces than those of gravity although the influence of gravity, acting always in one direction, would gradually have brought them towards the mean plane. In producing planets from a nebula we have to keep in mind the observed lifetimes of disks around young stars — one to a few million years. Every process in the bottom-up mechanism must take less time than a disk lifetime as must the total time for all the processes together. For an individual dust particle the settling time has been calculated to be of order ten million years, so this presents a problem.

A solution to the problem was suggested by the American planetary scientist, Stuart Weidenschilling, and his co-workers in 1989. It was suggested that there were a number of processes that would cause dust particles in space to stick together. To explain the first process it is first necessary to point out that, although the disks around stars have considerable mass, they are, by normal terrestrial standards, extremely diffuse. The dense part of a nebula disk would, on Earth, be designated as an ultra-high vacuum! Consequently the surfaces of the dust particles are not often impacted by gas molecules and they remain clean and uncontaminated. Laboratory experiments show that when two uncontaminated surfaces of the same material are pressed together the atoms on the surfaces can chemically bond. Such *cold welding* could occur between clean surfaces of dust particles if they came together.

The second reason why dust particles might join together is through electrostatic forces, due to particles becoming electrically

Figure 21.2 Computer simulation of the aggregation of dust particles.

charged by friction. This happens in everyday life with pieces of cling film or tissue paper that can be very difficult to detach from ones hands, or each other, due to the electrical charges they acquire.

A final reason that has been suggested is, paradoxically, the very opposite of the first one and considers that the dust grains might become covered with some sticky organic materials that would act like glue.

Weidenschilling and his colleagues did some computer simulations of how dust particles might stick together. One of the outcomes is shown in Figure 21.2.

The general form of these aggregates is that they are rather loose fluffy structures. The best way for the dust to come together from the point of view of fast settling would be to form very compact bodies but, according to theory developed by Weidenschilling, even the looser structures as shown in the figure would be able to settle in the mean plane on a timescale between one hundred thousand and one million years. This is an acceptable figure in the light of the observed lifetimes of disks.

Nevertheless, there is still some doubt about this result since Weidenschilling's calculations showed that the settling time is very sensitively dependent on the exact form of the dust aggregates. In 2000 a shuttle space experiment called CODAG (COsmic Dust AGgregation experiment) showed that dust in space did stick together quite readily but not in the form predicted by Weidenschilling's computer simulations. The aggregates were far more strung out in one direction forming a fluffy string rather than a fluffy blob. Because of this, and the sensitivity of the settling time to the form of the aggregates, there is some doubt that settling into the mean plane can take place within a disk lifetime. On the basis of the CODAG results, it was suggested that planetary scientists might have to review the way that planets formed but, later, those working on the CODAG project gave a less pessimistic picture, suggesting that as strings grew to a larger size they curled up, became more compact, and therefore could settle more quickly. The whole bottom-up process of planet formation depends on this first stage of dust settling. While opinions vary on how plausible it is, there are factors described in Section 32.2 that would make the settling process much more plausible.

Towards the end of 2005 an observation was made of an exoplanet in a triple star system designated as HD 88753. The planet was in orbit very close to a solar-type star. Two somewhat-less-massive stars, with combined mass 1.63 times that of the Sun, formed a close binary system in an orbit about the single star at an average distance of 12.3 au with eccentricity 0.5. This observation was accompanied by comment that it threw doubt on the bottom-up process of planet formation since the disk material accompanying the stellar system would be constantly stirred up, both thermally and mechanically, thus preventing the settling of dust into a carpet. A possible solution to this difficulty is that the planet formed in association with the solar-type star before the binary pair attached itself — although this explanation introduces new, if different, problems.

21.3 The Formation of Planetesimals

Once a dust carpet formed it would be a very substantial entity. The total mass of the disk could be up to one-tenth that of the Sun (Sun's mass = 2×10^{30} kg) and if one percent of that was dust then the mass of dust would be about that of Jupiter, 2×10^{27} kg. This mass, spread uniformly in a disk reaching out to the orbit of Neptune would give an average mass per unit area of about 30 kg m^{-2}. In 1973 it was shown theoretically by the American cosmogonists, William Ward and Peter Goldreich, that the dust disk would break up into a vast number of condensations, just as illustrated for the total disk in Figure 20.1. Because the dust disk would not be uniform, and also because of different distances from the Sun, the condensations would be of different sizes and masses in different locations. These condensations are the planetesimals, with typical dimensions in the range between hundreds of metres and tens of kilometres.

This part of the bottom-up process has a sound theoretical basis and it is fairly certain that if a dust carpet formed then planetesimals would be an outcome.

21.4 Making Terrestrial Planets and Cores for Giant Planets

Another theory that required planets to form from very diffuse material was the accretion theory, suggested by Otto Schmidt in 1944 and later developed by Lyttleton, as described in Chapter 12. Following this original idea, the theory of the first stage of forming planetesimals from dust was developed by two Russian workers but, because of the poor dissemination of information from Russian-language publications, this work was largely unnoticed outside the Soviet Union. The following process — that of building up larger bodies from planetesimals — was first tackled by Victor Safronov (1917–1999), who worked at the Institute of Earth Sciences in Moscow of which Schmidt was the Director. Safronov's work was published in Russian in 1969 but an English translation became available in 1972 and this idea soon became central in the development of the SNT.

When planetesimals first formed by fragmentation of the dust carpet they would have moved in close-to-circular orbits. Occasionally pairs of planetesimals would approach each other and through their mutual gravitational attractions they would be deflected out of their circular paths. Consequently, as time progressed, so the pattern of smooth circular motion would break down and the motions of the planetesimals would take on a somewhat chaotic character. As this

Figure 21.3 Victor Safronov.

happened, a new kind of mechanism increasingly came into play. The chaotically moving planetesimals sometimes collided and the effect of this was to damp down the chaos in the motion. Eventually a state of balance was reached between gravitational interactions, increasing chaotic motion, and collisions, decreasing chaotic motion, after which the overall degree of chaos in the motion remained more-or-less constant. Safronov showed that when this state was reached the relative speeds of planetesimals was about equal to the *escape speed* from the largest planetesimal.

The concept of escape speed plays an important role in many aspects of astronomy. If a ball is thrown into the air it comes down again but the faster it is thrown upwards the higher it goes before it returns. In the period after the Second World War, rockets that had been developed for military use were used as research tools by firing them into the upper reaches of the atmosphere to collect information beyond the reach of balloons. The faster the rocket was launched the further up it would go but there was a limit beyond which a single stage rocket could not go because of the relationship between the mass of its fuel and the energy it contained. Later, two and three-stage rockets were developed where the first stage provided a high platform for the firing of a smaller second stage, and so on. In this way rockets were launched that could escape the Earth's gravity and so become the carriers of space vehicles. For a spherical body of mass M and radius R the escape speed from its surface is given by

$$v_{esc} = \sqrt{\frac{2GM}{R}} \, . \qquad (21.1)$$

The escape speed from the Earth is about 11 km s^{-1}.

If two bodies move together and collide under the influence of their mutual gravitational attraction then, from theory, their relative speed on collision is at least equal to the escape speed of each from the other. If none of the energy of motion were lost then the bodies would simply bounce apart and separate again. However, if energy of motion is lost in the collision process, say in fragmenting one or both of the bodies or converting some energy into heat, then the two bodies can join together to form a single body. This process works better if one of the bodies is significantly bigger than the other when it can be pictured as accretion of the smaller body on to the larger (Figure 21.4).

The more massive a planetesimal is, the greater is its attractive force on other planetesimals and therefore the greater the number of other planetesimals it accretes. But, the more it accretes the greater

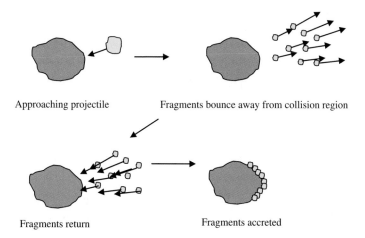

Approaching projectile   Fragments bounce away from collision region

Fragments return   Fragments accreted

Figure 21.4 The accretion of a small body by a larger body due to a collision.

is its mass and therefore the greater is its attractive force — and so on. The net result is that in any particular region of the disk one planetesimal will grow and become dominant. In the inner part of the Solar System these dominant planetesimals will become terrestrial planets but further out they will become the cores of major planets.

Safronov's analysis was mathematically based and he worked out the rate at which the cores or terrestrial planets grew and the total time for their formation. To a first approximation the time of formation of a body, t_{form}, is given by

$$t_{form} = \frac{r_B \rho_s P}{6\pi\sigma_{ad}} \tag{21.2}$$

where P is the orbital period of the body at the formation distance from the Sun, ρ_s, its density, σ_{ad} is the average areal density of planetesimal material in the formation region and r_B is the final radius of the body being formed. It has been estimated that to produce the Earth would have taken about one million years. Assuming that the disk mass is 0.1 solar mass, that σ_{ad} varies as $1/r$, where r is the distance from the Sun, and that a Jupiter core has 8 Earth masses and a Neptune core 4 Earth masses, then the time of formation of a Jupiter core is about 120 million years and of a Neptune core about 8 billion years. Changing the estimates of core masses and the variation of σ_{ad} with r can change these estimates somewhat but we now know that the time for producing Jupiter is unacceptable because of the observed lifetimes of disks, although that information was not available to Safronov when he carried out his analysis. The estimated time for producing Neptune is greater than the age of the Solar System, and Safronov *would have* been aware of that. This timescale problem was not ignored by the SNT community but it was thought that Safronov's scheme could somehow be modified to shorten the formation times.

21.5 Major Planets — The Final Stage

It must now be taken that planetary cores have been produced within a timescale such that a large quantity of nebula gas is still present. The Sun may have been much more luminous when it was first formed than it is now so that the temperatures in the region of the terrestrial planets may have been much higher than they are now. That fact, combined with the small masses of the terrestrial planets, may have prevented them from acquiring a gaseous envelope. Indeed, even at the present temperature of the Earth it would be unable to begin to acquire a hydrogen-plus-helium atmosphere.

In the outer part of the Solar System, from Jupiter outwards, conditions would have been suitable for the capture and retention of hydrogen and helium and calculations indicate that the acquisition of the gas, that is the majority component of Jupiter, would have taken no more than one hundred thousand years.

21.6 Some Problems with the Solar Nebula Theory

The bottom-up process starting with dust gave a complete model for the formation of planets backed up by some realistic analysis of the processes in each of the four stages. Some of the problems of the model were not known at the time it was first formulated and, as problems have become apparent, so work has been done to address them. However, a problem that was evident from the earliest development of the bottom-up process was the unacceptable time it would take to form the outer planets. An attempt to deal with this is described in Chapter 22.

Whatever the process by which planets are formed from a nebula, the expectation is that they would be orbiting in, or close to, the equatorial plane of the star. When highly inclined orbits were found for exoplanets, including retrograde orbits, the announcements of their discovery were accompanied by comments that they posed problems for nebula theories. Various mechanisms are being explored that might convert originally-direct orbits into ones that were retrograde. These include interactions of the retrograde planet with another, as yet undetected, planet that reversed its orbit or the influence of an external star that slowly modified the planetary orbit to make it retrograde — the so-called Kozai mechanism. Yet another possible, but rather involved, process, mediated by gravity waves within the central star, leaves the planetary orbits unchanged but modifies the rotation of the outer layers of the star so apparently changing the spin axis of the whole star.

So far the inclinations of very few exoplanets have been measured and the statistics are too uncertain to draw any deep conclusions. If, for example, it were found that a whole family of exoplanets was in the same highly-inclined orbital plane then the idea of planet-planet interactions could be ruled out. However, to echo the quotation that heads Chapter 18, it is probably better to have more data before going too deeply into theories of how directly-orbiting planets can have, or apparently have, their orbits reversed.

Chapter 22

Making Planets Faster

...things can never be done too fast.

Oliver Goldsmith (1730–1774),
She Stoops to Conquer

22.1 Conditions in the Disk

The process described by Safronov for making planets from planetesimals was certainly a positive step forward. If sufficient time were available then a mechanically based series of steps had been well described using sound physical principles and with a good mathematical analysis. At first it seemed that only Uranus and Neptune presented timescale problems but, once disk lifetimes had been estimated as a few million years at most, all the major planets were found to have formation times that were too long.

The American cosmogonist, George Wetherill (1925–2006), considered this problem in some detail. He proposed various conditions within the nebula disk that would substantially shorten the Safronov estimated formation times.

22.1.1 *The initial masses of planetesimals*

If the planetesimals in one region were all of similar size then it would take a long time for one of them to become dominant. By chance some would accrete more than others and, at first, a number of planetesimals would be competing for the role of 'dominant planetesimal'. Eventually just one would emerge and this would be able to collect much of the mass within its region of dominance. However, the process can be speeded up if the initial masses are very non-uniform, especially if one planetesimal is much more massive than anything else in its neighbourhood. This could come about if the nebula was somewhat 'clumpy' instead of having a smoothly varying density that just depended on distance from the Sun. A large clump would give rise to a large planetesimal and by this means a locally dominant planetesimal could be present straightaway.

22.1.2 *Big bodies move more slowly*

There is a principle in physics that goes under the name *the equipartition of energy*. It applies to a system of bodies, interacting with each other in any way, where not all bodies have the same mass. The type of interaction can be gravitational or by collisions — it matters not, the principle will still apply. Basically it says that the probable energy of motion will be the same for all bodies, regardless of mass. Since the energy of motion, known as kinetic energy, given by

$$E_K = \frac{1}{2}mV^2 \qquad (22.1)$$

increases with both mass, m, and speed, V, this implies that if the mass is higher the speed must be lower and *vice-versa*. This principle is illustrated in Figure 22.1.

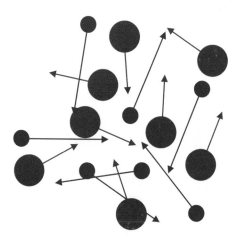

Figure 22.1   In an interacting system of bodies the more massive bodies move more slowly.

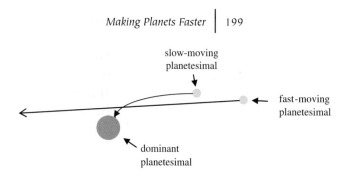

Figure 22.2 The gravitational deflection is greatest for slower moving planetesimals.

The application of this principle means that the dominant planetesimal will tend to be moving more slowly than the others and, of the remaining planetesimals, the larger ones will be the slowest movers. Hence, the most slowly moving planetesimals *relative to* the dominant one will tend to have greater than average masses. The importance of this is that the range of attraction for other bodies of a body with a particular mass increases when the relative speed is less. A slowly moving planetesimal will curl in towards the dominant body whereas a fast moving planetesimal will just pass by without much deflection (Figure 22.2).

For these reasons there will be a tendency for the dominant planetesimal preferentially to accrete the largest of the other planetesimals. This should increase the rate of growth over that calculated by Safronov.

22.1.3 *Gas drag*

The makers of motor vehicles and of aircraft are always concerned with the question of air resistance, or drag, the force on the vehicle due to its motion through the air. In the case of aircraft, flying at a great height reduces the density of the surrounding air and hence reduces the drag. This cannot be taken to the ultimate limit of leaving the atmosphere altogether as the presence of some air is required to give the lift that supports the plane. For a car there is no choice of environment; the car must travel through the air at ground level and all that one can do is to design its shape to offer as little resistance as possible.

The planetesimals would have been moving within the gas of the nebula and so will be subjected to drag that will slow down their motion. The drag will actually be more important for smaller planetesimals and any body that reaches planetary, or even moderate satellite, size will thereafter experience a negligible drag effect. It is all a question of how much surface is exposed to the gas, which for a given speed is the main factor determining the total force on the body, in relation to its mass. For two bodies of similar shape but differing in linear dimension by a factor of two, the larger body will have four times the area, and hence four times the resisting force, but this will be acting on eight times the mass — hence the slow-down due to drag on the body is less. Since gas drag will be relatively ineffective in slowing down the larger planetesimals it will be only marginally important for increasing the growth of the dominant planetesimal, but it will have some effect.

22.2 Runaway Growth

Applying all these factors led to what Wetherill has called *runaway growth*. It implies that the process by which the dominant planetesimal grows is an accelerating one so that the increase of mass gets faster and faster with time. The growth will slow down and cease altogether when all the material is locally exhausted. An analysis of runaway growth was described by G.W. Wetherill and G.R. Stewart in 1989.

The estimates of the times of formation of terrestrial planets and cores are drastically changed by runaway growth. A Jupiter core of ten Earth masses can now be formed in about one million years. The times for producing Uranus and Neptune are also much shorter, somewhere in the range from ten to thirty million years. The times for the outer planets are still uncomfortably large but at least they are within a factor of ten of being acceptable.

There are some serious doubts about the runaway growth model. An assumption in the model is that of a very high density of planetesimals, thus enhancing the rate of accretion by the dominant planetesimal. According to the American cosmogonist Alan Boss the assumption made by Wetherill and Stewart requires a disk density greater than can reasonably be expected. A further difficulty was raised by Wetherill himself. Once the dominant planetesimal has reached about one Earth mass then it will not only accrete planetesimals but it will also scatter them all over the Solar System. Once they are scattered they are not locally available to be accreted. While it could be argued that there will also be planetesimals scattered *into* the region of the dominant planetesimal these incomers will be moving at high speed and so be more difficult to accrete. Many more will be scattered than will be accreted so the local density of planetesimals that can readily be accreted will be much lower than if no scattering had taken place. The calculated rate of runaway growth assumes that all local planetesimals stay local and are always available to be accreted; the fact that they are not means that runaway growth will not be as fast as theory suggests.

By pushing a number of assumed parameters in favourable directions Wetherill and Stewart were able to reduce the estimated formation times of the outer planets. However, their analysis seems somewhat contrived, assuming factors specifically for the purpose of getting shorter times, rather than because such factors should be expected in any real situation. Despite their efforts, the timescale problem for forming the cores of the outer planets remained and other ideas were considered.

Chapter 23

Wandering Planets

Our souls, whose faculties can comprehend
The wondrous architecture of the world:
And measure every wand'ring planet's course,
Still climbing after knowledge infinite.

Christopher Marlowe (1564–1593)

23.1 The Need for Planets to Wander

Even if the basic ideas behind runaway growth were accepted, the outer planets, Uranus and Neptune, would still pose a time-of-formation problem. The difficulty stems from where they are formed. The further out they form the greater is the period of a planetary orbit and the lower is the space-averaged density of planetesimal material so, inevitably from (21.2), the Safronov theory is going to give very long formation times.

Another problem, knowledge of which is more recent, concerns the exoplanets that have been found very close to stars — as close as 0.04 au (Table 17.1). Assuming that these planets are gas giants, which seems likely from their masses, then it is improbable that they could have *formed* where they are now — although they are obviously stable in their present environments. A core in such a situation could not begin the process of building the planet by retaining local nebula gas that would be at a very high temperature.

A potential solution to both these problems is to posit that all the planets in question were produced in some more favourable position — say not too far from the present position of Jupiter (although this planet also has timescale problems) — where they can both form in a reasonable time and also acquire a gaseous envelope. Some could then move inwards to give the close-in exoplanets while others could move outwards to give planets at the distances of Uranus and Neptune from the Sun. Such movements are referred to as *migration* and processes for producing migration have been extensively studied.

One mechanism that can be directly discounted is drag due to the viscosity of the gas, which is totally ineffective for planetary masses, although it can have an effect on smaller planetesimals.

23.2 Interactions Between Planets

Although the SNT should produce planets in near-circular orbits they need not stay that way. Planets act gravitationally on each other and the orbit of a planet produced, say, in the outer region of the asteroid belt would be unstable over a long period of time due to perturbation by Jupiter. If its orbit became highly eccentric in this way then there could be a close passage of the two planets that could move one of them inwards towards the Sun while the other went out further from the Sun. If Jupiter were considerably more massive than the other planet then there could be a comparatively small inward change in its orbit while the other planet went out to a great distance (Figure 23.1).

This process alone could not explain the present orbits of Uranus and Neptune. The planet, say Neptune, may go out several tens of au from the Sun but its orbit would then be very eccentric and take it back into the vicinity of Jupiter's orbit. Some further process would be necessary for it to end up in a circular orbit of radius 30 au.

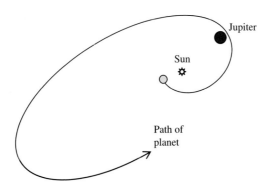

Figure 23.1 A planet thrown outwards by an interaction with Jupiter.

23.3 Effects Due to the Mass of the Nebula Disk

We put off a detailed description of the way that a planet interacts with a disk until Chapter 30 because there we consider the effects on general eccentric planetary orbits. For the present we just give a description of the evolution of the *circular* orbit of a planet within a disk. It is assumed here that the disk material circulates around the Sun (or a general star) in free orbit under the influence of the central mass so that at any distance from the Sun the material is moving in the same way as a planet would at that point. This means that there is little relative motion between a planet and the medium that is immediately adjacent to it.

If we consider material well inside, or well outside, the planet's orbit then there will be considerable relative motion. The inner material will be rotating faster and also moving at a higher speed than the planet. Conversely outer material will be moving at a lower speed than the planet, as indicated in Figure 23.2

Straightforward analysis (e.g. see reference to Cole and Woolfson, 2013, pp. 531–540) shows that the gravitational interaction between the planet and the slower moving outer material adds angular momentum to the material. The net effect of this is that the material moves further out and the planet moves further in. For the inner material, which moves faster than the planet, the reverse is true — the inner material loses angular momentum and moves inwards while the

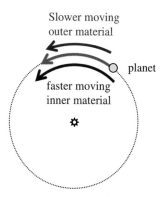

Figure 23.2 The relative motion of the planet with respect to neighbouring disk material.

angular momentum gained by the planet moves it outwards. The effect on the material is straightforward. It moves away from the planet whether it is inside or outside the planet's orbit. What is not quite so straightforward is the effect on the planet that gains angular momentum from inner material and loses it to outer material.

The movements of the medium away from the planet tend to open up a gap within the disk. When this occurs, the motion of the planet will be due to *Type II migration*, first described by the American astronomer William Ward in 1997. Whether the planet moves inwards or outwards depends on the balance of the effects from the inner and outer material. Inward migration takes place far more readily and can take the planet very close to the central star, as is observed for some exoplanets. Outward migration is much more difficult to achieve and, when it does take place, the total outward movement is small. Type II migration is not a realistic mechanism to transfer planets formed near Jupiter or Saturn out to the orbit of Neptune.

For a planet with mass less than about that of Jupiter no gap appears so the planet is always in contact with the medium. This gives rise to what is called *Type I migration*. This is caused by the asymmetric effect of wakes forming on either side of the planet that have the appearance of spiral waves. These are akin to the straight wakes produced by a moving boat; in the present case the effect of the circular motion of planet and medium is to distort the wakes into a spiral form. The outer wake tends to push the planet inwards and the inner wake tends to push it outwards but the net effect always leads to inward motion and is extremely effective. Indeed, it is so effective that it seems that almost any planet involved in Type I migration must plunge into the central star. A mechanism that will prevent this from happening will be described in Section 23.5.

Figure 23.3 shows the state of the medium under the influence of a planet in a circular orbit found from a simple computer model. Gaps can be seen in the medium although not a clean circular gap as the material is very mobile. Other features worth noting are the spiral structures both inside and outside the planet's orbit.

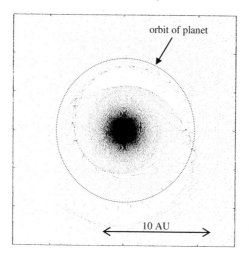
orbit of planet

10 AU

Figure 23.3 The disturbance of the medium by a planet in a circular orbit.

23.4 The Role of Spiral Waves

The spiral waves seen in Figure 23.3 are carriers of angular momentum. Those moving outwards from the planet's orbit are transporting angular momentum from the region of the planet outwards. As they travel they dissipate, meaning that they give up energy and angular momentum to the medium within which they move. Another way they can dissipate is by impinging on some external planet that then acquires energy and angular momentum. If an outer planet is constantly being fed by these waves generated by a massive interior planet then this is a way of moving the outer planet outwards. Angular momentum is conserved in the system by inward motion of the planet that produced the spiral wave.

It has been suggested that Uranus and Neptune were moved outwards by such a process, the provider of the spiral waves being Jupiter. Since Jupiter has ten times the mass of Uranus and Neptune combined, a comparatively modest inward motion of Jupiter can give up sufficient angular momentum. This is an interesting idea but it has not been convincingly demonstrated that it will work. An important consideration is the efficiency of the process — as seen in Figure 23.3 the spiral waves are extensive, of the order of many au in extent, while the planets are very small targets for the absorption of angular momentum and energy. For this reason it seems doubtful that they can be greatly affected by this process, especially as there is limited amount of time (a few million years at most) available for the mechanism to operate.

In Chapter 17 the account of the observation of dust disks around older stars and the direct imaging of exoplanets showed the presence of planets at much greater distances from stars than Neptune is from the Sun — more than 100 au. Their locations present a challenge to the SNT even more difficult than that of the outer Solar System planets.

23.5 Saving the Planet

When a planet is moving inwards towards the star, under either Type I or Type II migration, its motion is in the form of a shallow spiral. At any particular time it is moving closely in the form of a circular orbit around the central star. Once it gets close to the star the tidal forces can become large enough to give significant distortion to each body. Here we are just concerned about the distortion of the star.

The mechanism that is to be described depends on two periods, that of the planet in its orbit and that of the star's spin where the latter is to be smaller than the former. To fix our minds we take a star of solar mass with a planet in a circular orbit of radius 0.05 au. Such an orbit has a period of about four days and we now take the spin period of the star as three days. Such a star is spinning about nine times faster than the Sun but that is not unreasonable for a young star before its spin has been modified by the processes described in Section 19.5.

As the planet orbits the star so the tidal bulge on the star will tend to face the planet. However, the faster spin of the star drags the bulge in a forward direction; the situation is illustrated in Figure 23.4.

The force on the planet due to the bulk of the star points towards the centre of the star and neither adds or subtracts from the angular momentum and energy of the planet's orbit. The force due to the

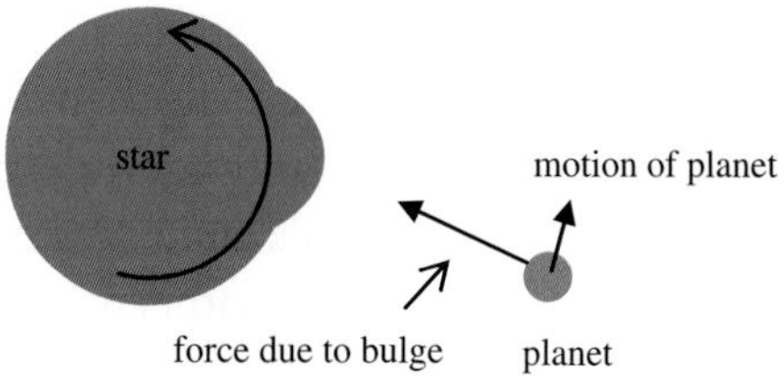

Figure 23.4 Forces on an orbiting planet due to the tidal bulge on a rapidly spinning star.

bulge is mostly towards the centre but also has a small component in the direction of the planet's motion. This will add both angular momentum and energy to the motion of the planet and so will tend to move it outwards. This gain can just balance the loss of angular momentum and energy due to motion in the disk so the planet's orbit is then stabilized and it will go no closer to the star. This is quite a stable situation. If the planet were to move inwards then the tidal effect would dominate and push it out again. Conversely if it happened to move outwards then the medium resistance force would dominate and move it in again. The planet would be saved!

23.6 A Problem with the Terrestrial Planets

The Type I migration mechanism is very effective — so effective that it gives timescale problems for the formation of the terrestrial planets, a matter raised by the Japanese astronomers E. Kokubo and S. Ida in 1998. When a terrestrial planet, or a planetary core, is forming by the accumulation of planetesimals the rate of growth of its mass is proportional to $m^{2/3}$, where m is the current mass. Hence the growth is slow at the beginning and then accelerates as the core grows and becomes more efficient in attracting further planetesimals. Small bodies, like planetesimals or cores in their early stage of growth, are strongly affected by Type I migration and would spiral from the terrestrial region into the Sun on a timescale much shorter than the time required to grow the terrestrial planets. They would have insufficient mass to raise a tide on the star of sufficient size for the mechanism in Section 23.5 to operate.

It has been suggested that this problem can be resolved by considering that the terrestrial planets were formed after the gas disk had dispersed by planetesimals that had migrated into the terrestrial region when the gas was present. While the cores of gas giants must be formed while the disk is present — otherwise they could never become gas giants — there is no such requirement for the terrestrial planets.

Chapter 24

Back to Top-Down

There is always room at the top.

Daniel Webster (1782–1852)

24.1 Perceived Problems with the SNT

Although the SNT offers mechanisms for almost all aspects of forming planets around stars, it has constantly faced timescale and other problems. After almost four decades of development, by the end of the twentieth century the mechanisms that underlie it had been theoretically described but not convincingly demonstrated. The timescales for producing major planets by planetesimal aggregation was the dominant problem but there were others such as the uncertainty that dust could settle into a plane to enable the planetesimals to form — an essential first step — although in Chapter 32 a possible solution to that difficulty is proposed. Another important problem, that of retrograde exoplanet orbits, was still over the horizon. Given the real and apparent difficulties of the standard bottom-up process for producing planets, the American cosmogonist, Alan Boss (b. 1951), returned to a top-down approach but one involving principles not previously considered.

24.2 The Rotating Disk Model

In 2000, Boss numerically modelled the rotation of a viscous gas disk. The normal circular orbiting speed for material varies with distance from the centre of the disk and this set up friction between neighbouring layers of gas that eventually created a rather uneven structure including the presence of filaments of material of density much higher than the local average. A representation of the

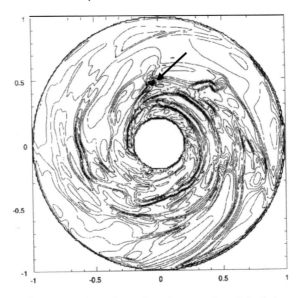

Figure 24.1 Filaments and condensations in a rotating disk of viscous material.

configuration of the material at a particular time, as given by Boss, is shown in Figure 24.1.

Within the disk there is a condensation, marked with an arrow, that has a mass about five times that of Jupiter. However, it did not persist and disappeared after about sixty years of simulated time. There were a number of assumptions in the first model produced by Boss that merit comment. The first of these relates to the temperature of the material. Normally, when gas is compressed it heats up; those who have pumped up a bicycle tyre will know that constant compression of the gas makes the pump quite warm. Boss assumed that the gas did not heat up when compressed, an assumption that favoured the formation of condensations. However, in later calculations Boss showed that if the material cools quickly when heated — a quite reasonable assumption — then the results are not very different from those without heating.

Another point of doubt is the assumed parameters of the disk. It had a mass of 0.091 $M_\odot$, below 0.1 $M_\odot$ usually taken as the maximum for a disk, and with radius 20 au, which is the minimum value suggested for circumstellar disks by Beckwith and Sargent in 1996.

One might expect *a priori* that a comparatively massive disk would be at the upper end of the expected size range rather than at the low end. This combination of nearly extreme, and somewhat conflicting, assumptions increased the density in the disk and hence the possibility of producing a condensation. Again, although it was unknown at the time the model was first suggested, since it produces planets within a circumstellar disk it faces the same problem of retrograde orbits as the other models of planet formation in a disk, and must depend on the same proposed mechanisms for resolving the problem (see Section 21.6).

In 2008, Boss published a paper in which he investigated the sensitivity of his model to the resolution of his modelling technique. By using a higher resolution he produced condensations with a longer lifetime than those of his original model, and he suggested that even higher resolution would lead to stable condensations.

Despite the reservations about the assumed parameters of the disk, the approach is interesting and brings a new feature into the process of planet formation. A difficulty of the basic SNT is that it starts with material in a very diffuse form and requires a multi-stage bottom-up process to produce planets; at least one of these stages, the aggregation of planetesimals, has immense, possibly insuperable, timescale problems. In the Boss model, some of this diffuse material is concentrated into higher-density filaments within which the well-understood phenomenon of gravitational instability operated to produce a condensation — albeit one that was short-lived. If the filament material had been denser and the line density of the filament had been greater, as Jeans postulated for his tidally-produced filament, then a permanent condensation would have been produced. However, although Boss had pushed the parameters of his model to the limit it failed to produce what was required. It did, however, get close.

In Section 21.2 it was mentioned that the presence of an exoplanet round a solar-type star in the triple star system HD 88753 had been given as a reason to doubt that a dust carpet could form, due to the turbulence that would be caused in the disk. If the results in Figure 24.1 are valid then it would seem that turbulence would occur in the disk even for an isolated star. While turbulence would be helpful to the Lynden-Bell and Pringle angular momentum transfer mechanism described in Section 19.2, it introduces yet another uncertainty about the viability of the process of forming a dust carpet.

Part VII
Making Stars

Chapter 25

This is the Stuff that Stars are Made Of

...an infinite deal of nothing...

William Shakespeare (1564–16516),
The Merchant of Venice

25.1 The Question

15 gm butter
3 eggs
a pinch of salt

Lightly whisk the eggs in a bowl and then melt the butter in a non-stick pan. Strain in the eggs plus salt and return to a brisk heat. As the mixture cooks at the edge lift up the edges to allow liquid egg to flow out from underneath. Rock the pan to-and-fro to avoid sticking. When liquid egg ceases to appear from underneath, tip out the contents of the pan onto a plate so that it is neatly folded across the centre.

What we have given above is a recipe for making an omelette. It consists of two parts, the first giving the ingredients and the second describing the process by which the ingredients are transformed into the final product. So it is for the formation of astronomical objects — including stars. Here we shall be mainly concerned with the first part, the ingredients, and we ask the question, "What are stars made of?"

A good way to seek an answer to this question is to use powerful telescopes to explore the galaxy and thus to see what possible ingredients there are within it.

25.2 The Galaxy

In the search for stellar ingredients there is no point in looking outside our galaxy because galaxies are well-separated entities within

217

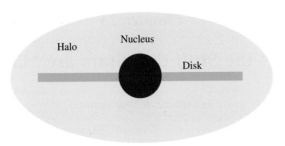

Figure 25.1 A schematic view of the Milky Way galaxy seen edge-on to the disk.

each of which star formation is taking place. Since we are going to explore the galaxy with our telescope this is a good opportunity to describe the galaxy that we live in — the Milky Way. It has a fairly simple structure. The spiral arms, seen in the similar galaxy NGC 6744 (Figure 2.3), occur within a disk-shaped region. The Sun is situated within the Milky Way disk about one-third of the way out from the centre. Around the centre of the disk there is a spherical densely populated region, called the *nucleus* of the galaxy, containing many stars. Finally, there is a lightly populated region, the *halo*, in the shape of a flattened sphere, stretching beyond the boundary of the disk (Figure 25.1).

So what do we see in our telescopes when we explore the galaxy? First we see large numbers of individual stars of various kinds, many of them similar to the Sun. Actually, some are seen to be double stars where two stars move around each other. There are other, apparently single, stars that can be inferred to be double stars by various kinds of physical measurement — by a wobble in their motions or from Doppler shift observations that indicate that the star we see is actually two unresolved stars. These double-star systems are called *binary stars* and the interesting conclusion from observations is that the number of binary systems is roughly equal to the number of single stars. Thus, only one-third of all stars are single stars like our Sun.

Another type of stellar system revealed by our telescopes is clusters of stars and they are of two main kinds. The first kind normally contains a few hundred stars that are separated well enough for the individual stars to be easily seen. They only occur in the disk of the galaxy for which reason they are referred to as *galactic clusters* although, alternatively, in view of their diffuse nature, they are also referred to as *open clusters*. A typical galactic cluster is shown in Figure 25.2.

The second type of cluster consists of several hundred thousand stars and they are called *globular clusters* (Figure 25.3). They may occur anywhere in the galaxy, including the halo. They are spherical in form and in the central regions individual stars are not readily resolved.

The fact that so many stars are seen in clusters suggests something about the way that stars are formed. Indeed, many astronomers

Figure 25.2   The Pleiades, an open cluster.

Figure 25.3   The globular cluster M13.

believe that *all* stars, including single stars like the Sun as well as binary stars, were originally members of clusters. Through gravitational interactions with other stars in the cluster, individual stars may sometimes acquire enough speed to be able to escape from the cluster; calculations suggest that after one hundred million years or so a galactic cluster will have completely evaporated. Indeed, the term *evaporation* is very appropriate since a pool of water evaporates by individual water molecules acquiring enough speed to escape from the surface, and eventually the whole pool disappears in this way.

25.3 The Ingredients

But now let us return to the question of ingredients. Clearly, the best place to detect the ingredients for making stars is in a region where stars are being made and we can locate such regions. One such is the Orion Nebula (Figure 25.4), a region of gas and dust within which young bright stars and also newly-forming cool stars are observed. This gas and dust is obviously what the stars are being made of and there are several other star-forming regions rich in gas and dust. However, a moment's thought leads us to wonder about the origin of these clouds of dust and gas — are they really the ultimate source,

Figure 25.4 The Orion nebula — a star forming region. It is illuminated by four very young bright stars contained within it.

the root source? We swing our telescope hither and thither and we see nothing else. Distant stars are seen, bright and clear and with nothing impeding the passage of light. The space between stars and outside the star-forming clouds is apparently empty.

But wait. Although we can see stars at great distances, when we observe carefully we find that the further away a star is the redder is the light from it. It needs accurate observations to confirm this effect but there can be no doubt about it. From theory we know that tiny dust particles could produce such reddening but they must be very thinly spread if the light from stars tens of thousands of light years away can reach us without being blocked out. In fact, these dust particles are part of what is called the *interstellar medium*, usually referred to as the ISM, which occupies all the space between stars in the galaxy. The dust that causes the reddening of light is only about one percent by mass of the ISM, the rest being gas, mostly hydrogen. Let us create a picture of how diffuse the ISM really is. Look at the last joint of your little finger. That volume in the ISM contains one or two hydrogen atoms. That same volume in the air around you contains about 10^{20} nitrogen and oxygen atoms. The particles of dust are very tiny; about ten thousand of them would fit in the dot above the letter *i*. Now imagine a cube with a side of one kilometre. In such a cube within the ISM there is *one* particle of dust! In actual fact, the ISM is not completely uniform but is somewhat lumpy. In many regions it will have the composition as just described and a temperature of about 7,000 K. In other regions it may be more or less dense with, respectively, somewhat lower or higher temperatures.

Ultimately this is the stuff that stars are made of — the ISM — and we shall have to see by what process these ingredients, in such an unlikely form, can be made into stars.

Chapter 26

Making Dense Cool Clouds

...and maketh the cloud...

Book of Common Prayer

26.1 The ISM, Clouds and Temperature

It is clear from observations that star formation takes place within clouds of gas and dust that are much denser than the ISM. They are also much cooler and, typically, a dense cloud of density 10^{-18} kg m^{-3} will be about one thousand times as dense as the ISM and have a temperature about 10 K, a small fraction of the ISM temperature. In the passage from the very diffuse ISM to the production of a dense star (the average density of the Sun is 1.4 times that of water) the formation of a dense cool cloud (DCC) is an essential first step; the combination of high density and low temperature give a Jeans critical mass similar to that of a galactic cluster whereas the critical mass for the ISM is similar to that of a globular cluster. Here we shall describe a mechanism by which the conversion of the ISM to a DCC can take place.

When we examined the galaxy with a big telescope, as described in Chapter 25, what we could *not* detect is that energy is constantly traversing the space between stars. One obvious energy source is starlight, without which we would be unable to see the stars. This is generally very weak but in regions near a very bright star it could be quite significant. The other source of energy is *cosmic rays*. These are not really *rays* in the sense of being a form of radiation but rather they are charged particles moving at speeds very close to the speed of light and so possessing a great deal of energy. They permeate the whole of the galaxy and come from all directions. One problem occupying the attention of astronomers is that of the source of cosmic rays but, whatever their source, of their existence there is no doubt.

A characteristic of this energy traversing the galaxy is that it is absorbed by matter and, when that happens, the matter is heated. This will be so for both the matter that comprises the ISM and also that of the DCCs. Since the ISM and DCCs do not steadily heat up then clearly there must be some cooling process at work that exactly compensates for the heating. One simple form of cooling is through the radiation of heat by the dust particles. This is exactly similar to the loss of heat from a central-heating radiator — the hotter it is the more heat it radiates. However, there are other important modes of cooling, but to discuss these we need to know more about the nature of the material that comprises the ISM and DCCs.

26.2 Atoms, Ions, Molecules and Electrons

ISM and DCC material consists of atoms, ions, molecules and electrons. Virtually all the mass of an *atom* is contained in its nucleus. It contains protons, with positive charges, and neutrons, of virtually the same mass as a proton but with no charge. Around the nucleus there are electrons, equal in number to the protons, each with a charge of the same magnitude as that of a proton but opposite in sign. Thus, the atom as a whole has no charge since positive and negative charges cancel out. The overall diameter of an atom is typically 10^{-10} m so that one million of them, side by side, would span the dot over a letter *i*. Most of the volume of an atom is occupied by the electrons; if the size of the nucleus is represented by a fist then the electrons are several kilometres away! A schematic representation of a carbon atom is given in Figure 26.1.

Molecules consist of a number of atoms bonded together, a well-known example being where two hydrogen atoms plus one oxygen atom join together to form a molecule of water (Figure 26.2).

Sometimes one or more electrons can be knocked out of an atom. The released electrons, called *free electrons,* can then move around independently leaving behind an *ion* with a net positive charge, as shown in Figure 26.3.

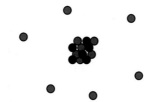

Figure 26.1 A representation of a carbon atom (not to scale). The nucleus contains 6 protons (red) and 6 neutrons (black). The six electrons (blue) exist in a comparatively large region around the nucleus.

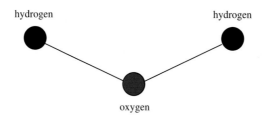

Figure 26.2 A water molecule.

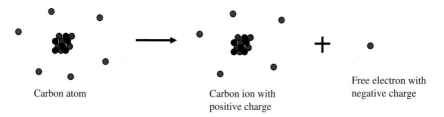

Figure 26.3 Ionization of a carbon atom to give a carbon ion plus a free electron.

26.3 Further Cooling Processes

As explained in Section 2.1, temperature is a measure of the energy of motion of the particles of which the material consists — for the ISM and DCCs these particles are atoms, molecules, ions and electrons. Due to their very low mass and the equipartition principle described in Section 22.1.2, for a given energy (temperature) electrons move much faster than the other particles and they are constantly colliding with atoms, molecules and ions. These collisions are responsible for a number of cooling processes but we shall explain just one of them here.

A branch of modern physics, called quantum mechanics, tells us that the electrons in atoms and ions can only exist in states with certain allowed energies. They can jump, or be pushed, from one energy state to another, either up or down, but what they cannot do is exist with an energy that is not one of those allowed. We now consider the collision of a free electron with an atom (or ion), illustrated in Figure 26.4. One of the electrons in the atom can be pushed into an allowed state of higher energy while the free electron correspondingly loses energy of motion. The electron in the atom prefers to return to its original state with a lower energy and it does so. The energy it gives up in this process is converted into a packet of radiation (a *photon* of usually visible or ultra-violet light), which leaves

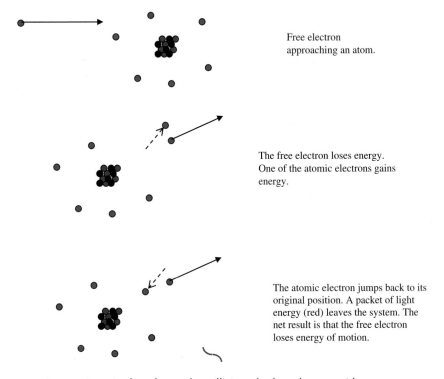

Free electron
approaching an atom.

The free electron loses energy.
One of the atomic electrons gains
energy.

The atomic electron jumps back to its
original position. A packet of light
energy (red) leaves the system. The
net result is that the free electron
loses energy of motion.

Figure 26.4   Cooling due to the collision of a free electron with an atom.

the neighbourhood with the speed of light. The net effect is that the original free electron, and consequently the material as a whole, has lost energy of motion and, hence, has reduced in temperature.

There are other processes, some involving molecules, which give cooling. An important feature of all these cooling processes is that they give a greater cooling rate when the material density is higher, because with denser material the rate of free-electron, and other, collisions per particle is increased. Another controlling factor is the temperature itself with the obvious relationship that, in general, the higher the temperature the greater the speed of electrons and other particles — and, hence, the rate of collisions — and the greater the cooling rate. Knowledge about the various cooling processes that operate in the ISM and DCCs goes back a long way. The British atomic physicist, Michael Seaton (1923–2007), gave the first theory of cooling by the excitation of atoms and ions in 1955 and Chushiro Hayashi (1920–2010), a Japanese astrophysicist, analysed the role of dust cooling in 1966.

Although cooling rates depend on both density and temperature, heating rates, especially by cosmic rays, are very little affected by the state of the material. To a first approximation, the heating rate, taken as the energy absorbed per unit mass of material per unit time, can be taken as constant regardless of the form of the material.

26.4 Making a Dense Cool Cloud

To explain the next step in the process of forming stars, illustrated in Figure 26.5, we must assume that some stars already exist. When massive stars get old they reach a stage where they suddenly explode very violently. This kind of event is called a *supernova*. The pressure waves from the explosion compress the local ISM, so increasing its density, and in addition a great deal of rather dusty debris is injected into the surrounding space. We have seen that when the density of material is increased, the cooling rate also increases and this is what happens to the compressed region of the ISM. Another enhancement of the cooling rate is provided by the extra dust in the region. The heating rate is unchanged but the cooling rate increases

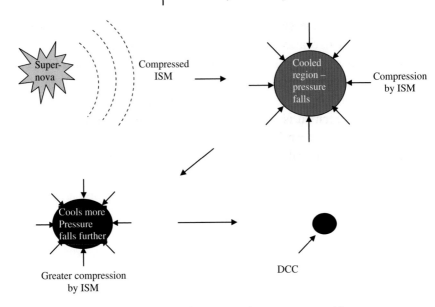

Figure 26.5 The stages in the formation of a DCC, triggered by a supernova.

thus leading to a fall in temperature. Now another effect comes into play. If the temperature goes down then so does the pressure — pressure, P, depends on the product of density, ρ, and temperature, T, according to,

$$P = \frac{\rho k T}{\mu} \qquad (26.1)$$

where k is the Boltzmann constant and μ is the mean molecular mass of the gas. Although the local density has increased slightly, the fall in temperature is such that the net effect is a fall in pressure. Now the external pressure of the ISM is greater than that within the cooled region so the ISM squeezes the cooled region, compressing it even further and increasing its density. But this further increases the cooling rate and the whole process continues, with the density rising and the temperature falling in the cooled region. We are forming a DCC. However, the process eventually comes to an end because, although the cooling rate increases with increasing density, it also decreases with decreasing temperature and eventually the influence of the low temperature overcomes that of increasing density.

The way that the system adjusts itself is that, at the end of the process, the DCC that has formed is cooling at the same rate as it is heating so that it stays at a constant temperature. In addition the pressure within it is similar to that of the ISM so there is no tendency for it either to be compressed by, or to expand into, the ISM. The DCC that has formed is stable, or nearly so, and we have taken the first step towards forming stars. The computational modelling of this process was described by Yann Golansky and myself in 2001.

Chapter 27

A Star is Born, Lives and Dies

Twinkle, twinkle little star,
How I wonder what you are,
Up above the world so high,
Like a diamond in the sky!

Jane Taylor (1783–1824)

27.1 Collapse of Stout Party

When a DCC forms it is at a low temperature and high density and is in approximate pressure equilibrium with the ISM. What we now consider is the size, and hence the mass, of the cloud. For a typical DCC, with a temperature of 10 K and density 10^{-18} kg m^{-3} (1,000 times that of the ISM) the Jeans critical mass, given by Equation (11.2), is about 1,000 solar masses, so that if the DCC mass exceeded that value it would begin to collapse.

A general DCC would have an irregular shape, although the calculation of the critical mass by Jeans was on the basis of a spherical cloud. It is reasonable to take the cloud as spherical for theoretical purposes, as this simplifies calculations about its behaviour, but we have to accept that the numbers we get from our calculations are just indicative of general values in a more realistic non-spherical case and should not be taken too literally. In this vein we now consider the way that a spherical DCC with mass greater than the Jeans critical mass would collapse under gravity.

If the cloud material started from rest we might conclude after observing it for a short time that nothing much was happening. What we would be witnessing is what is called *free-fall collapse*. It starts very slowly, gradually gets faster and eventually, in the final

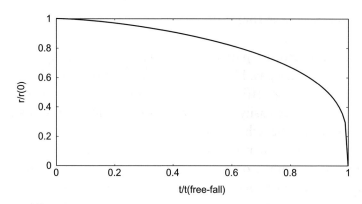

Figure 27.1 The free-fall collapse of a sphere. The times (*t*) are fractions of the time for complete collapse and radii are fractions of the initial radius.

stages, is very rapid indeed. Figure 27.1 shows how the radius of the spherical cloud changes with time, slowly at first and then ever more rapidly. Although free-fall collapse does not strictly apply to gaseous material, because it only takes account of gravity forces and ignores pressure effects, it describes the initial stages of the collapse of a cloud reasonably well, especially if the mass of the cloud appreciably exceeds the critical mass.

The free-fall time for the total collapse of a uniform sphere was worked out by James Jeans and is given by,

$$t_{ff} = \sqrt{\frac{3\pi}{32\rho_0 G}} \tag{27.1}$$

where ρ_0 is the initial density of the material and G the gravitational constant as described for Equation (11.1). For a DCC of density 10^{-18} kg m^{-3} this would be just over two million years.

To produce a star of solar mass, 2×10^{30} kg, it is necessary to have a density that will make the left-hand side of Equation (11.2) at least equal to the right-hand side, and preferably greater. For a temperature of, say, 20 K and material with mean molecular mass 4×10^{-27} kg (a mixture of hydrogen, molecular hydrogen and helium) the critical density is just under 10^{-14} kg m^{-3}. This density is 10,000 times greater than the density of the newly-formed cloud so clearly the cloud, or some part of it, must collapse further before a star can be formed.

When a cloud, or part of it, collapses then it heats up due to compression of the gas. However, the cooling processes, described in Chapter 26, are very efficient and since radiation can pass out of the diffuse cloud quite easily, its temperature remains low. Considering the collapse of the whole cloud, if its entire mass fell smoothly towards the centre then it might be expected that one very massive star, containing much of the original mass of the cloud, would be produced. But that is neither what we want nor what actually happens so we must now consider the process that, instead, produces a large number of normal-mass stars.

27.2 Turbulent Times

In a type of film shown in the early days of the silent cinema one might have seen the heroine adrift in a canoe on a wide slowly-flowing river. All is well but suddenly the river begins to flow more quickly, although still smoothly. We sense that trouble is ahead. The river is getting narrower and, as it does so, the flow gets faster. Now the river enters a narrow gorge, the flow is not only fast but it is turbulent with the water violently crashing around. Streams of water travelling in opposite directions crash into one another and splatter water high into the air. The canoe is pummelled this way and that. It is time for our hero to take a hand.

What we have described in this scene is a transition from smooth, called *streamlined*, flow into *turbulent flow*, and this transition takes place for any fluid, liquid or gas if it goes beyond some critical speed. Coming back to our free-falling collapsing gas sphere the motion of the gas is slow at first but then becomes ever faster. Eventually it can no longer fall inwards in a smooth streamlined way but becomes turbulent. Superimposed on an overall collapse of the sphere there are gas streams moving randomly with respect to each other and occasionally colliding, a situation illustrated schematically in Figure 27.2. We shall now see how it is that these colliding streams of gas can give us individual stars.

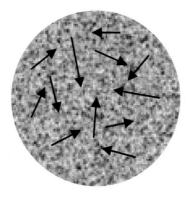

Figure 27.2 A collapsing cloud with turbulence. The general direction of motion of the material (shown by the arrows) is inwards but there is considerable turbulence present.

27.3 The Big Squeeze

When the streams of water collided in the ravine they splattered high into the air. Water is incompressible and the streams of water could not go through each other so they had to change direction. Gasses are different. When they collide they can be compressed and a clump of gas can form with a density higher than that of the oncoming colliding streams (Figure 27.3).

Here and there in the turbulent collapsing cloud colliding gas streams produce high-density clumps. The act of compressing the gas also heats it and the mass of the compressed clump will almost certainly be less than the Jeans critical mass for its initial density and temperature. However, two things will happen to the clump after it forms. The first is that the gas will begin to re-expand and the second is that the gas will cool, because of the cooling mechanisms described in Chapter 26. The critical factor here is that cooling is a much faster process than re-expansion. Before very long there is a clump of gas, very little less dense than in its original compressed state but much cooler, and the conditions may then be such that the Jeans critical mass is exceeded. The clump will have a stellar mass and it will begin to collapse. A star is born!

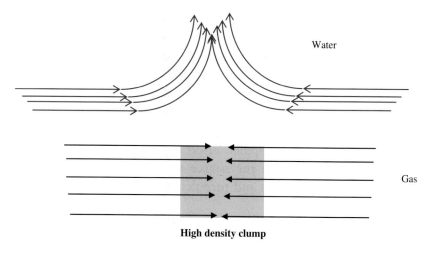

Figure 27.3 The behaviour of streams of water and streams of gas when they collide head-on.

27.4 Some Observations About Star Formation

Turbulence in star-forming regions is an observed phenomenon. Such regions are sources of *maser* (acronym for *microwave amplification by the stimulated emission of radiation*) radiation — like the light from a laser but of much longer wavelength. The source of such radiation at particular characteristic frequencies can be associated with various chemical entities, such as water. The water molecule shown in Figure 26.2 can vibrate in various ways but the rules of quantum mechanics apply, so the energies of a particular form of vibration can only have discrete values. The energy loss in jumping from an allowed vibrational state of higher energy to one of lower energy gives a packet of electromagnetic energy, corresponding to a wavelength in the radio part of the spectrum. What is true for water is true for many other molecules and free radicals[1] and from the

---

[1] A free radical is a bonded group of atoms that is unstable and would chemically react with other matter under terrestrial conditions but has a long lifetime in the rarefied conditions of the ISM or a DCC. An example of a free radical is the hydroxyl group OH.

wavelengths observed the nature of the emitting material can be determined.

The mechanism that produces the maser action, that makes the sources so bright, is not well understood, but it exists and is easily observable against a low background of microwave radiation covering a wide continuous range of frequencies. Observations from star-forming regions show that there are shifts in the characteristic frequencies due to the Doppler Effect (Section 17.2) that indicate random radial motions of the source material. These motions are associated with turbulence and turbulent speeds are estimated as of order 10 km s^{-1}.

For a very young galactic cluster the new stars are not on the main sequence but moving towards it on the tracks shown in Figure 16.3. If many stars in a cluster have luminosities and temperatures corresponding to the region of these tracks then this both indicates that the cluster is indeed young and also enables the masses and ages of the new stars to be found. In 1969, the British astronomers Iwan Williams and William Cremin examined data for a number of young stellar clusters and were able to deduce certain patterns in the way that stars were formed. Three main conclusions can be drawn from their study, which are, in qualitative form:

(i) The rate at which stars form increases with time.

(ii) The first stars produced are, on average, somewhat above solar mass and subsequently there are two streams of development, in one of which the average mass of stars increases with time and in the other of which the average mass of stars decreases with time.

(iii) The number of stars produced per unit mass range decreases with increasing mass.

27.5 A Star-Forming Model

In 1979 I investigated a model of a collapsing DCC. The analytical and numerical modelling gave quite good agreement with observations. The analysis showed that a collapsing cloud would inevitably become turbulent. As the cloud collapsed so gravitational energy was released that turned into energy of motion — some giving over-all inward motion and some appearing in the form of turbulence, as illustrated in Figure 27.2.

In the modelling of the collision of turbulent elements it had to be taken into account that not every collision would produce a star. First of all the density of the material must be fairly close, within a factor of four or so, to that satisfying the Jeans minimum mass for a star in order that a small amount of compression will raise the density to the right level to give collapse. If the streams collide too gently then the degree of compression will be insufficient but, con-versely, if they collide too violently the material simply gets splat-tered and will not form a star. The need for a narrow range of collision conditions regulates the rate at which stars are produced and the model gave results closely agreeing with the Williams and Cremin observations — including the feature of an increasing rate of production with time. The first stars produced in the modelling were of 1.35 solar masses and subsequently had less mass; after some time stars that had been produced earlier moved through dense parts of the cloud and were able to accrete material to form a small number of more massive stars. The number of stars per unit mass range found in the numerical results also agreed well with the conclusions from the Williams and Cremin study.

It has been observed that stars with masses less than about 1.35 times that of the Sun tend to spin slowly, like the Sun, while more massive stars have higher rates of spin. The speed of material moving round the equator of the Sun is 2 km s^{-1} whereas for a star with ten times the Sun's mass the equatorial speed is typically 200 km s^{-1}. The stars produced directly by collisions in my model, all with masses less than 1.35 solar masses, were shown to have low rates of spin (a few times greater than the present value for the Sun but capable of being brought down to their present values by the mechanism described in Section 19.5). However, for more massive stars, pro-duced by further accretion, the spin rates from the model were found to be much higher. In fact, the results from the model were in remarkable agreement with observations of spin rates for stars of different masses.

Actually, my original model for the formation of more massive stars, by the accretion of gas on to an existing star, has been thrown into doubt by subsequent work. Ian Bonnell, a British astronomer, and his colleagues, have shown that the radiation from a massive star, is so intense that unattached cloud material will be driven away from it rather than being accreted. They have therefore looked at a more plausible model in which the coalescence of lower mass protostars forms massive stars. Supporting the Bonnell conclusion, the American astronomer Mark Krumholz and his colleagues, based on supercomputer numerical modelling, showed that once a star forms it cannot accrete much extra mass in the form of unattached cloud material. Actually, the formation of massive stars by the Bonnell process, involving the addition of low-mass protostars coming from random directions to build up a large mass, would also lead to rapid rotation rates, similar to what was deduced from having material added in a more nebulous form.

27.6 Binary Star Formation

It has already been noted that the number of binary-star systems is approximately equal to the number of individual field stars, like the Sun. There are two quite distinct types of binary star system. In the first of these, the two component stars can both be resolved; such a system is called a *visual binary*. However, these are comparatively rare and are far outnumbered by *spectroscopic binaries*, where the stars are so close to each other that they cannot be resolved by telescopic observation so that the only way of determining that two stars are present is through the Doppler Effect.

When a protostar is formed the angular momentum it contains can vary over a considerable range. Let us consider a spherical protostar with a radius of 3×10^{14} m (2,000 au) and density 2×10^{-14} kg m^{-3}, giving a mass about 1.13 times that of the Sun. The galaxy is spinning at an angular speed of 10^{-15} radians[2] per second and we assume

---

[2] A radian is a natural unit of angle and is the angle subtended at the centre of a circle by an arc of the circle equal in length to the radius. One radian is 57.29°.

that the protostar is spinning at 10^{-14} radians per second — a modest value, just ten times the rate of the galaxy. If the protostar shrank to stellar dimensions, with a final radius of order 10^8 m, then, neglecting any change in the distribution of matter in the star, according to Equation (9.1) in order to conserve angular momentum the angular speed would be,

$$\omega_{final} = \omega_{initial}\left(\frac{\text{original radius}}{\text{final radius}}\right)^2 = 10^{-14}\left(\frac{3\times10^{14}}{10^8}\right)^2$$
$$= 0.09 \text{ radians per second}.$$

This would give a star with a spin period of 70 seconds and with equatorial material moving at 9,000 km s^{-1}; such a star could never form since it would fly apart under centrifugal forces. That being so, we ask ourselves how such a protostar *would* evolve.

The form of the collapse of a body under gravitational forces depends on the distribution of matter in the body and the nature of the material. The form of behaviour that would be expected in this case is illustrated in Figure 27.4.[3] Starting with an approximately spherical distribution (Figure 27.4a), collapse gives an increase in rotational speed, causing a bulge at the equator and change of shape to an *oblate spheroid*, rather like a flattened football. A general ellipsoid is a three-dimensional shape for which projections down directions x, y and z are all different ellipses. Figure 27.5 shows a general ellipsoid for which the semi-axes are a, b and c. An oblate spheroid is when $a = b > c$. The next stage, shown in Figure 27.4c, is a general ellipsoid, known as a *Jacobi ellipsoid* in this context, in which the ratio of the three axes of the ellipsoid changes until $a{:}b{:}c = 1.89{:}0.81{:}0.65$ after which the body takes on a pear shape (Figure 27.4d). Finally, a neck appears between the two ends and eventually the body breaks up into two parts, one much more massive than the other, which orbit around each other (Figure 27.4e). Each of the parts collapses further to form the component stars of a binary system.

---

[3] A more detailed account will be found in M.M. Woolfson, *On the Origin of Planets; By Means of Natural Simple Processes*, 2011 (London: Imperial College Press).

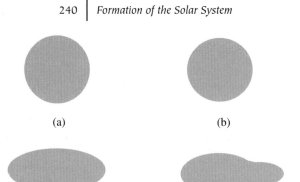

(a)

(b)

(c)

(d)

(e)

Figure 27.4 The stages in the simulated collapse of a spinning liquid sphere. (a) The original sphere; (b) An oblate spheroid; (c) A Jacobi ellipsoid; (d) A pear-shaped configuration; (e) Fission into two bodies. Between (a) and (e) the scale is reduced by a factor of about 1,000.

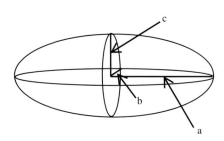

Figure 27.5 A general ellipsoid. The semi-axes are a, b and c.

Although, for clarity, the different parts of Figure 27.4 are shown at about the same size, the system is collapsing and the scale is actually changing considerably from one part of the figure to the next. The sphere shown in Figure 27.4a has a radius of 2,000 au but the component parts of Figure 27.4e may be separated by only about 2 au. In some circumstances, if the initial angular velocity is large enough, the smaller of the two blobs in Figure 27.4e may end up with such a high speed that is completely escapes from the larger blob thus giving two individual stars.

The separation of the two stars formed by this process is normally small enough to give a spectroscopic binary although, rarely, the separation may be large enough to form a visual binary. A more common way of forming a visual binary system is when the individual stars come together to form a bound system, but this requires special circumstances that will be described in Section 28.3.

27.7 The Death of a Star

While there are still many uncertainties in the mechanics of star formation, there is little dispute that protostars form and about the way that protostars evolve. When stellar condensations are first produced they are nothing like the stars we see in the sky — the Sun, for example. They are large cool spheres of gas — protostars. A protostar with one half the mass of the Sun at a density of 10^{-14} kg m^{-3} has a radius of just under 2,000 au — more than 60 times the radius of Neptune's orbit. Their long journey before they become normal main-sequence stars like the Sun was described in Section 16.2. For a star like the Sun, the complement of hydrogen fuel gives a lifetime of ten thousand million years after which it evolves to the stage of becoming a *red giant* star, with radius about 1 au. After shedding a great deal of outer material it ends up as a *white dwarf*, which, typically, has the mass of the Sun in a body the size of the Earth.

More massive stars have a shorter lifetime, down to a few million years and in their evolution after the main-sequence stage they generate internal temperatures that enable nuclear reactions to occur generating ever-heavier elements with the heaviest being iron. Eventually, the pressure in the core becomes so great that the protons and electrons in atoms combine to form neutrons. When this happens the core rapidly collapses and incoming material, falling inwards to occupy the resulting empty space, collides with outwardly moving material to give a catastrophic explosive event called a supernova. The end product of such an event is either a *neutron star* or, for the most massive stars, a *black hole*. But the occurrence of a supernova is where we began our story about making stars.

Part VIII
Capture

Chapter 28

Close to the Madding Crowd

Far from the madding crowd's ignoble strife

Thomas Gray (1716–1771)

28.1 Neighbours

The Sun is a solitary *field star,* which means that it makes its way through the galaxy and is not a member of a cluster or a binary system. By measuring the distances of stars in the vicinity of the Sun it is found that, locally, the average distance between stars is about 7 ly (reminder: a light-year, ly, is the distance travelled by light in one year and is about 10^{13} km). The closest star to the Sun is Proxima Centauri at a distance of 4.3 ly. We found out that the space between the stars was filled by the ISM and it is remarkable that this very diffuse medium, which is quite difficult to detect, accounts for almost as much mass as the stars in the solar neighbourhood.

In Section 25.2 we described the two types of cluster — galactic (or open) clusters typically containing a few hundred stars and globular clusters containing a few hundred thousand stars. Globular clusters are not of interest in our story since the stars they contain are very old, unlike the Sun. They contain comparatively little in the way of elements heavier than hydrogen and helium, the two lightest elements. Stars like the Sun are found in galactic clusters and such stars contain one percent or so of heavier elements. It is a common assumption that the Sun began its existence in a galactic cluster from which it eventually escaped — although there is another and more likely possibility, mentioned later.

We recognize the existence of a cluster because in the field of view of a telescope stars are seen to be clumped together. Clusters vary enormously in the numbers of stars they contain and their dimensions but, characteristically, the average distance between stars in a galactic cluster is about 1 ly. As clusters are three-dimensional structures it means that the number density of stars (number per unit volume) in a cluster is some 350 ($\sim 7 \times 7 \times 7$) times that in the solar vicinity. This still makes the stars a long distance apart. Taking the Solar System, as defined by the planets, to have a radius of 30 au the distance between stars in a galactic cluster is more than 1,000 solar-system diameters.

28.2 Another Big Squeeze

We now return to the collapsing turbulent DCC described in Chapter 27, with protostars being produced here and there as streams of gas collide in a suitable way. As the cloud collapses so its average density increases and its temperature will also tend to increase, although the temperature rise will be greatly moderated by the cooling processes described in Chapter 26. Although we describe the star-forming region as 'dense', that is only by astronomical standards — it is actually quite diffuse and transparent to radiation, so cooling processes will be quite efficient. Since the density is increasing substantially and temperature is rising little, according to Equation (11.2) the Jeans critical mass for the cloud material will consequently decrease and stars of smaller mass will tend to be produced as time progresses. Of course, the conditions in the cloud will be highly variable; for one thing it will certainly be denser and hotter in its core than in its outer regions. At any time, stars with a range of masses will be produced, depending on where and how they originate, but the average trend will be for stars with progressively smaller masses to form, which is what is deduced from observations. Observations also show that more massive stars are being produced as the cluster develops. It is consistent with the ideas of Bonnell and his colleagues (Section 27.5) that massive stars occur towards the centre of clusters where there is a high density of lower mass protostars, which gives favourable conditions for aggregation.

The protostars being produced in the collapsing cloud become compact objects, YSOs as described in Section 16.2, in a few free-fall times corresponding to their original density — say, in 50,000 years — a small fraction of the free-fall time for the DCC as a whole, which is of the order of two million years. The new stars are involved in the general inward collapse so that not only the density of the cloud material, but also the *number density* of stars (i.e., the number of stars per unit volume) within it, increases with time. Clusters of stars being produced within dense clouds are known as *embedded clusters* and the cluster is gravitationally bound together by the gravitational influence of the cloud material within which it is embedded. In the 1990s it was found from infrared observations, which can detect low-temperature sources, that most young stars were found in dense embedded clusters, where the number density of stars, at all stages of development, could be anything up to 3,000 per cubic light year.

The dense embedded stage lasts about five million years, after which some of the more massive stars undergo a supernova event, the energy from which drives out the gas that binds the cluster together, thus causing the cluster to expand. In ten percent of cases the expansion eventually ceases and gives an open cluster but in the other ninety percent of cases the stars completely disperse to become field stars or isolated binary systems. Hence, it is more likely that the Sun was part of a cluster that dispersed than that it was a long-time member of a galactic cluster from which it eventually escaped.

28.3 Interactions in a Dense Embedded Cluster

During the dense embedded stage the average distance between stars can be as small as 0.1 ly or even less. Protostars are still produced during the embedded stage so what we have are protostars, with radii about 2,000 au, being formed in an environment where compact stars are about 0.1 ly (6,500 au) apart. An impression of the relative dimensions of this situation is given in Figure 28.1. The protostars will stay as big diffuse objects for a few thousand years; the free-fall collapse time for a protostar of initial density 10^{-14} kg m^{-3} is, from (27.1), more than 20,000 years. It is clear that, given the relative motion of stars and protostars, interactions between bodies in a dense embedded cluster must occur, especially interactions involving the largest bodies,

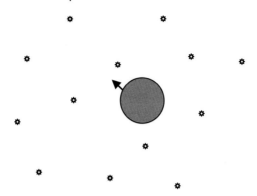

Figure 28.1 An impression of a protostar in the presence of compact stars within an embedded cluster.

protostars. The formation of large mass stars by the accumulation of numbers of protostars, as suggested by Bonnell and his colleagues, is most likely to take place during the dense embedded stage of a cluster's development.

We have seen that a large proportion of stars occur as binaries, and the way that binaries could form was described in Section 27.6. Some binaries, where the two stars are close together, are very tightly bound — in the language of astronomers they are *hard binaries* — whereas others form a much looser combination, so-called *soft binaries*. The passage of a star close to a soft binary can cause its disruption and this is something that could happen in a dense embedded cluster. The Czech, later Australian, astronomer, Pavel Kroupa (b. 1963), has studied the frequency of binary systems in dense stellar environments. In particular he determined the frequency of binaries in two star-forming regions, the Taurus-Auriga dark cloud and the Orion Nebula. The number density of stars is much higher and the frequency of binary systems much lower in the Orion Nebula. These observations suggest that stellar interactions with binary systems, leading to their disruption, have occurred more frequently in the Orion Nebula — a consequence of it having a higher number density of stars.

We now recall the Jeans idea of two tidally-interacting stars (Chapter 11). A diffuse protostar would be very greatly affected if it came under the influence of strong tidal forces. So the question arises, "Will an appreciable number of protostars in a diffuse form pass close to condensed stars — YSOs or main-sequence stars — and, if they do so, what will happen to them?"

Chapter 29

Close Encounters of the Stellar Kind

Yet meet we shall, and part, ...

Samuel Butler (1835–1902)

29.1 Jeans Revisited — The Capture Theory

The Jeans tidal theory, described in Chapter 11, had proved to be untenable but, in developing it, Jeans produced sound and unchallenged analysis concerning the formation of a filament due to the tidal disruption of a star and how condensations would form in such a filament due to gravitational instability. The two major criticisms that led to the demise of the Jeans tidal theory were as follows:

1. Material from the Sun was too hot to form a planet.
2. Material could not be pulled sufficiently far from the Sun to explain the planetary orbits.

In 1962, I investigated a model based on a tidal interaction between the Sun and a diffuse body of mass 0.15 times that of the Sun with radius 14 au. Such a body could be interpreted as a protostar in its final collapse stage. The scale of this early model was chosen with an eye on the size of the Solar System so the nearest approach of the protostar to the Sun was about 44 au. In this model it is the diffuse body, the protostar, which will be distorted and form a filament in which condensations would form.

The idea being investigated, which gave the essential difference between this model and that of Jeans, was to see whether the filament condensations, considered as protoplanets, could be *captured by the Sun* — thus giving the name *Capture Theory* (CT) to the proposed mechanism. Fortunately, this idea arose when computers were improving rapidly and I had access to the most powerful computer in the world at that time, the Ferranti Atlas. Its capabilities would now be regarded as derisory — although it cost over two million pounds, its capacity and speed were a small fraction of what is now comfortably carried in a pocket or handbag. Nevertheless, it was good enough to test the capture-theory idea with a simple model and the results were published in 1964. The behaviour of the protostar is illustrated in Figure 29.1, which shows the result of the computation.

Because of computer limitations, and the difficulty of obtaining access to a computer in those days, the model was a very simple one, even to the extent of representing the protostar as a two-dimensional object. The protostar was modelled as a distribution of mass points, initially placed on a grid, with forces between points representing the effects of gravity, pressure and viscosity. Also present was the gravitational force of the Sun. The model was far too crude to show condensations forming in the filament but, since Jeans had a good analytical model for this, various points along the filament were selected as potential condensation centres. Computation showed that some of the material represented by these points went into elliptical orbits that were within the region around the Sun occupied by the planets.

This model is free of the objections that had been raised about Jeans' original theory. Condensations in the filament, derived from the protostar, could collapse to form planets since the protostar was a cool body. Again, since the protoplanet blobs came from the protostar they did not have to be pulled out of, and then move well away from, the Sun and the scale of their orbits was primarily determined by the path of the protostar relative to the Sun.

29.2 New Knowledge — New Ideas

In the light of current knowledge the original version of the CT is quite untenable — although it did show the possibility of applying Jeans' elegant theory in ways that avoided the difficulties of the Jeans tidal model. A protostar of radius 14 au would have been collapsing quite quickly and therefore the chance of it passing a compact star while it was still in an extended state would have been

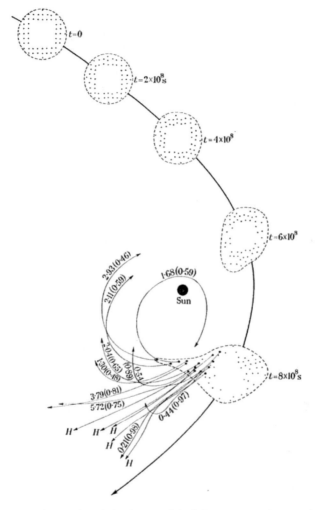

Figure 29.1 The results of the first model of the Capture Theory. The final orbits marked *H* are hyperbolic meaning that they left the protostar but were not captured by the Sun. For other orbits the perihelion is given, in units of 10^{12} m, and also the eccentricity (from Woolfson, 1964).

very small. It was also assumed at the time that the environment in which the tidal interaction took place was within a galactic cluster, with stars having an average separation of 1 ly and moving with an average speed of 30 km s^{-1}. It was concluded from these conditions that one star in 10^5 would be expected to have planetary companions — which could not be ruled out at that time because the only planetary system known was the Solar System.

Over the more-than half century since the CT was first proposed, there has been a tremendous growth of new knowledge, based on observations, which any theory of the origin of the Solar System must take into account. As knowledge grew so the CT was adapted to accommodate it and it was encouraging that, although the conditions under which it was applied were required to change, particularly involving large changes in the scale of the model, the basic capture mechanism was robust enough to accommodate all the changes without itself needing to change.

There were two main observations that required a radical rethink of the conditions under which the CT was applied. The detection of other planetary systems and the realisation that they were quite common (Chapter 17) was the first of these. The second was the establishment of the existence of a dense embedded stage in the evolution of a galactic cluster. These led to a reformulation of the conditions under which the capture-theory mechanism operated that included the idea of taking the protostars as being close to their newly formed state, with initial radii of order hundreds up to 2,000 au, in which they would remain without much collapse for a considerable time. We shall see that, under these conditions, although the form of the CT interaction is as previously described, the behaviour of the protostar is somewhat different from that previously found.

29.3 A Method for Realistic Simulations

A very effective technique used in astrophysical simulations is called smoothed-particle hydrodynamics (SPH). In SPH, introduced independently in 1977 by the American astronomer L.B. Lucy and by the Anglo-Australian partnership G.A. Gingold and J.J. Monaghan, the astronomical object is represented by a set of points each of which represents some of the object's mass and also carries with it other properties of the material, such as velocity and thermal energy. There are various forces between the points that realistically simulate gravitation, gas pressure and the viscosity of the stellar material.

The original 1964 modelling of the CT could be considered as a very crude forerunner of SPH. During the application of SPH the temperature associated with the points changes, due to either compression or expansion of the material and also to the conversion of kinetic energy of motion into heat energy through the agency of viscosity.

An important piece of physics that was not included in the original 1977 formulation of SPH is that of radiative heat transfer. Heated objects radiate heat energy that is either absorbed by other bodies or lost from the system. Over the years various approximations had been used to take radiative heat transfer into account. At one extreme, if the bodies concerned were very transparent to radiation then it was assumed that they took on the temperature of the local environment — that is to say they stayed at the same temperature regardless of expansion, compression or the action of viscosity. At the opposite extreme, if the bodies were extremely opaque to radiation, so that the passage of radiation through them is slow, then it was assumed that temperature changes due to heat transfer by radiation can be ignored by comparison with those due to dynamic effects. Other approximations have been made for intermediate conditions that usually involve taking the temperature of the material as some function of its density. However, for this particular application — modelling the CT — a rather better approach was required. Different parts of the model were very different in their characteristics in terms of size and opacity and the central star, irradiating everything in the system, removed any possible relationship between temperature and density. For this reason, in 2003 Stephen Oxley and I devised a new procedure for simulating radiation transfer that closely mimicked the actual physical process that occurs. The whole region of the simulation was divided into smaller regions each of which was a source of radiation, emitted in the form of packets of energy, rather like the packets of energy that physicists call photons, which have already been referred to in Section 26.3. These energy packets were either absorbed by other small regions, the temperature of which changed, or they passed out of the system. In the context of radiation transfer the central star was a constant source of radiated energy. The procedure was tested under conditions where the outcome could be predicted and was found to work well, its only drawback being that it greatly increased the computational burden.

29.4 Capture-Theory Simulations

In the simulation of a CT event there were various parameters that needed to be set. There were:

(i) The mass of the central star (M_*).
(ii) The luminosity of the central star (L_*).
(iii) The mass of the protostar (M_P).
(iv) The radius of the protostar (R_P).
(v) The temperature of the protostar (T_P).
(vi) The composition of the protostar indicated by its mean molecular mass, μ.
(vii) The initial distance of the protostar from the star, D.
(viii) The initial motion of the protostar. This was defined by fixing the closest approach of its centre to the star (q) and the eccentricity of its orbit relative to the star, (e).
(ix) The number of SPH points representing the protostar (N).

In Figure 29.2 there is shown a simulation, from a 2004 study by Oxley and myself with the following parameters:

$$M_* = 2\times10^{30}\,\mathrm{kg} = M_\odot \quad L_* = 4\times10^{26}\,\mathrm{W} \quad M_P = 7\times10^{29}\,\mathrm{kg} = 0.35\,M_\odot$$
$$R_P = 800\ \mathrm{au} \qquad\qquad T_P = 20\ \mathrm{K} \qquad \mu = 4\times10^{27}\ \mathrm{kg}$$
$$D = 1{,}600\ \mathrm{au} \qquad\qquad q = 600\ \mathrm{au} \qquad e = 0.95$$
$$N = 5{,}946$$

The figure shows the SPH points projected onto the plane of the motion. Initially, the protostar is spherical but after 6,000 years it has approached the star and has been greatly distorted. After a further 6,000 years the whole protostar has been stretched out into a filament, which is not the behaviour described by Jeans or shown in Figure 29.1. However, as was expected from the work of Jeans, finally the filament breaks up into a series of condensations. It was very gratifying to see the confirmation of Jeans' analysis on gravitational instability illustrated in such a graphic way. These condensations eventually collapse to form potential *protoplanets*.

Five of the condensations shown in the last image of Figure 29.2 are captured into orbit about the star. These captured condensations have masses (in Jupiter units), and orbits with semi-major axes and eccentricities given by,

(4.7 M_J, 1,247 au, 0.835), (7.0 M_J, 1,885 au, 0.772), (4.8 M_J, 1,509 au, 0.765), (6.6 M_J, 1,325 au, 0.726), (20.5 M_J, 2686 au, 0.902).

The first four masses are higher than most, but not all, masses observed for exoplanets (see Table 17.1). However, Figure 29.3

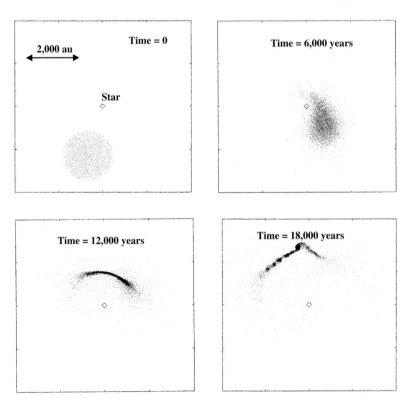

Figure 29.2 A sequence of profiles for a capture-theory SPH simulation at 6,000 year intervals.

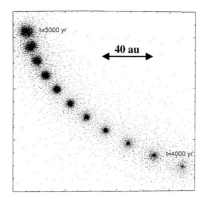

Figure 29.3 The collapse of a protoplanet as it goes into orbit around the Sun, shown at 100 year intervals between 3,000 and 4,000 years after the origin of the condensation. At the end of the period it consists of a central collapsing core surrounded by a disk of material.

shows the collapse of a protoplanet from another SPH simulation over a simulated period of 1,000 years where the total mass is about five times the mass of Jupiter. It was found that in the final stages just over one half of the mass is in the central core so that the mass of the final planet is less than that of the original condensation. The remainder of the mass is in the form of a surrounding disk.

The final captured body has a mass such that it would not be a planet at all but rather a *brown dwarf*. These are bodies that span the gap between planets and stars. They have the characteristic that they become hot enough to ignite nuclear reactions involving deuterium, an isotope of hydrogen (see Chapter 40), but never hot enough to ignite hydrogen and hence become a main-sequence star. As a reasonable guide we may take any body with a mass in the range 13–70 times the mass of Jupiter as a brown dwarf. Above that range the body is a star — below that range a planet. It should be noted that in the search for exoplanets several of the stellar companions have been found to have masses in the brown-dwarf range.

Some of the planetary-mass condensations produced in the filament are not captured by the star but end up as independent bodies in the forming stellar cluster, eventually to become bodies within interstellar space when the cluster evaporates. These bodies relate to observations made by the British astronomers, Philip Lucas and Patrick Roche, in 2000. Using infrared observations they were searching for brown dwarfs in the Orion Nebula — which they found — but they also found 13 bodies in the planetary mass range that they described as *free-floating planets*.

A characteristic of the captured protoplanets that needs to be further considered is their orbits, which are much more extensive than those that have been observed either in the Solar System or for exoplanets. Indeed, there is no possibility that orbits that have been observed could be produced directly by the capture-theory mechanism as described here.

29.5 Doing without Protostars

A possible problem with the simulation shown in Figure 29.3, and others like it, is that they start with the protostar in the form of a sphere. At the initial distance from the star this is unrealistic. If the protostar formed in that position by a suitable collision of turbulent elements of the cloud then it would have originally have been distorted; if it had formed at greater distance then by the time it reached the starting distance for the simulation it would also have been distorted. Actually, from the point of view of producing planets, starting with a spherical protostar is a *disadvantage*. Since the whole process depends on the protostar being stretched into a filament, starting with a sphere rather than an elongated shape hinders, or at least delays, the process.

Work on star formation, described in Section 27.5, indicated that to form a protostar the conditions for the collision were somewhat restricted — it had to be neither too gentle nor too violent. In the rather turbulent conditions that prevail in a cloud it is violent collisions that usually occur and the only way to produce a protostar is if the turbulent elements approached each other rather obliquely so that their mutual relative approach speed was reduced. In our 2004 paper, Oxley and I investigated what would happen to the compressed material produced by fairly violent collisions of streams in the vicinity of a star — something that would happen far more frequently than the interaction of a star with a protostar. In this case, to describe the interaction it is necessary to fix the mass of each stream, their initial positions, velocities, shapes and densities as well as other parameters. Figure 29.4 shows the result of a collision between two similar streams, each of mass 0.5 $M_\odot$ and of cylindrical form. A number of protoplanet condensations are produced, three of which, marked with letters, are captured with mass, in Jupiter units and orbits with semi-major axis and eccentricity as follows:

A (1.00 M_J, 4,867 au, 0.768), B (1.6 M_J, 1,703 au, 0.381),
C (0.75 M_J, 1,736 au, 0.818).

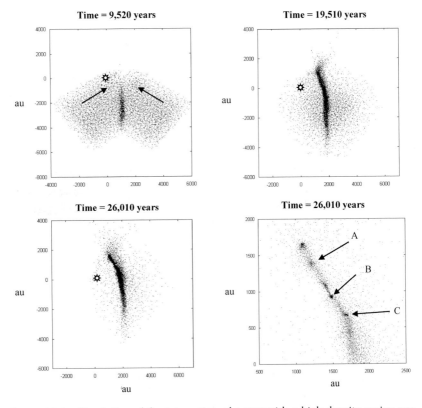

Figure 29.4 Simulation of the interaction of a star with a high density region produced by colliding turbulent gas streams at three times — 9,520 years, 19,510 years and 26,010 years. The final image shows a higher resolution view of part of the filament at 26,010 years.

In this case, condensations with masses less than or equal to that of Jupiter are formed but, again, the problem of modifying the orbits to correspond to what is observed requires attention.

As has been mentioned in relation to the evolution of the CT, the mechanism is extremely robust in the sense that parameters can be varied over very wide ranges to give simulations that produce captured planets. If protostars and high density regions produced by the collision of turbulent elements *are* being produced in an embedded cluster in the presence of compact stars, a scenario supported by observations, then it seems that capture-theory events would inevitably take place. The question is, "With what frequency?", and that is what will be considered in Chapter 31.

29.6 Some General Comments on Planet Formation

In Section 8.2 one of the conditions given for a plausible theory of the origin of the Solar System, and of planetary systems in general, is that is must be able to produce planets. While this may seem an obvious condition — so obvious that it is almost unworthy of mention — it turns out that the majority of theories put forward cannot meet this most basic requirement. The problems of planet formation for the SNT by the bottom-up approach have been described in Chapter 21, with attempts to solve the problems in subsequent chapters. The underlying problem is that, although there is plenty of material to form planets in a circumstellar disk it is in such a diffuse form that it is difficult to assemble it by a bottom-up approach in a reasonable time into the compact form representing a protoplanet. The mechanism described by Boss in Chapter 24 has come closest to forming a planet, starting with disk material, but only because the mechanical effect of having a spinning disk concentrated the diffuse material into filaments — although it turned out that the density in the filaments was not quite enough convincingly to give a permanent condensation, at least as far as present modelling has shown.

By contrast with the SNT mechanism, the CT has as its source of planetary material a protostar with mass a few times that of a maximum-mass disk but with initial density considerable less than that expected within a disk. However, the tidal mechanism that draws the whole protostar into a filament so concentrates the material that stable protoplanetary condensations readily form. What is clear, both from the Boss model and from the CT simulations, is that the most likely mechanism to produce protoplanets is through the gravitational instability of a dense filament, as envisaged by James Jeans one hundred years ago.

Chapter 30

Ever Decreasing Circles

The energies of our system will decay,...

A.J. Balfour (1848–1930)

30.1 The Initial Orbits of Protoplanets

The protoplanets produced in the calculations illustrated in Chapter 29 were on orbits very unlike those either of planets in the Solar System or of any observed orbits of exoplanets. The initial orbits of the protoplanets were ellipses with semi-major axes of order 1,000 au or more and eccentricities around 0.9 while, for comparison, Neptune, the outermost solar-system planet, has an almost circular orbit of radius 30 au. The comparison is illustrated in Figure 30.1.

For the CT mechanism of planet production to be plausible it is necessary for there to be some mechanism for producing the required decay and rounding-off of orbits from the initial states to final states corresponding to what is observed in the Solar System and for exoplanets.

30.2 A Resisting Medium

The formation of protoplanets, as illustrated in Figures 29.2 and 29.4, not only gave rise to captured and free protoplanets but also to a substantial amount of material captured by the star and forming a resisting medium, in the structure of a disk, around it. A body, such as a planet, moving in a resisting medium would be expected to lose energy, leading to an orbit of ever decreasing size. An example of this at a smaller scale is what happens to artificial Earth satellites in low orbits. Although the atmosphere is very thin at heights of tens

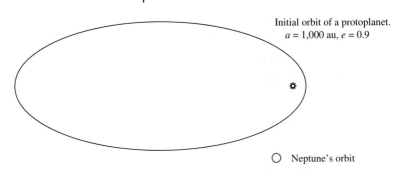

Figure 30.1 A comparison of an initial capture-theory protoplanetary orbit and that of Neptune.

of kilometres, it gradually causes the decay of the satellite's orbit, so bringing it closer to the Earth. As it approaches the Earth the atmosphere becomes denser and the rate of decay increases. The final stages that lead to the satellite plunging to Earth are quite rapid and not easy to predict precisely so there is always concern about where the satellite will land — the sea, or somewhere in an unpopulated land area being desirable.

The resistance of a fluid to the motion of a solid body within it is a well-known phenomenon in everyday life. Try moving your hand quickly in water — the resistance of the water limits the speed with which you can move it. What is more, the faster you move your hand the greater is the resistance you feel. This is a characteristic of the way that resistance operates. An object that is stationary with respect to a resisting medium experiences no force at all. When it moves within the medium it experiences a force, and the greater the speed of the object relative to the medium the greater is the force. Obviously, the direction of the force is such that it opposes the motion of the object.

Although these statements indicate the general way that resistance operates it turns out that there are several different mechanisms through which resistance can occur. One of these applies to objects of any kind moving in any kind of fluid medium, where the force exerted depends on the nature of the medium, the speed with which the object moves and the size and shape of the moving object, but not its mass. In another type of mechanism the gravitational effect of the mass of the moving object acting on the medium plays the dominant role and, finally, there is a mechanism where the gravitational action of the object on the medium and that of the medium on the object are both important.

30.3 Resistance Due to Viscous Drag

It is a matter of everyday experience that water flows easily, that treacle flows less readily and pitch, that becomes fluid on a hot day, flows extremely sluggishly. The property of these materials that leads to their different rates of flow is their *viscosity*, a form of internal friction that inhibits the relative motion of neighbouring fluid layers. If a body moves through a fluid it drags the fluid with it and so causes different parts of the fluid to move with different speeds. Viscosity forces within the fluid react back on the object causing the motion and constitute a resisting force on the moving object.

In Figure 30.2 we see the flow of air around a streamlined object. The flow is smooth and consequently the viscosity force on the object is small. If the object were a rather blunt shape, for example, a cube, then the flow would be less smooth and there would probably be turbulence (see Section 27.2). Turbulence means that neighbouring bits of fluid are moving rapidly with respect to each other and this will generate big viscosity forces and hence high resistance to the motion of the body.

An important characteristic of viscous resistance is that for bodies of similar shape and density the more massive the body the less it is affected. The reason for this is that the force on the body varies with

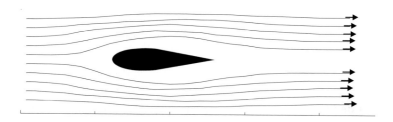

Figure 30.2 Motion of a fluid around a streamlined object.

the area impacting on the fluid, which varies as the square of the linear dimension, but this force acts on a mass that is proportional to the cube of the linear dimension. Thus, the greater the size of the body the less is the force per unit mass, or deceleration. For large bodies, such as planets or satellites, forces due to viscosity are insignificant compared to the other forces that operate.

30.4 Resistance Due to the Effects of Mass

Let us suppose that the resisting medium through which the body moves does not consist of a fluid but is composed of a large number of more-or-less uniformly spaced solid objects that are so far apart that they do not collide or come close to each other. In this case there is no viscosity of a conventional kind. The solid objects act on each other through gravity but these forces are just related to the separation of the bodies and not their relative motion.

Now let us imagine that a body moves through a region occupied by these solid objects, which are initially at rest, without touching any of them. Due to the mass of the body, after it has passed through the region the objects will be moving — i.e., they will possess energy that they did not have originally. This energy has to come from somewhere. Some of it may come from changes in gravitational potential energy due to the rearrangement of the solid objects but the rest comes from the body that loses energy by slowing down. The slowdown of the body's motion is the result of it experiencing a resisting force. This force has nothing to do with viscosity and theory shows that, in many situations, the rate of deceleration of the body is proportional to the mass of the body. What we have deduced for a medium consisting of solid objects would also apply to a fluid, e.g., a gaseous, resisting medium that would also acquire energy from the passage of the body. The density of the medium plays a part in this resistance mechanism in that the greater is the density of the medium the greater the energy it gains and, hence, the greater the resistance.

Finally, for a very dense and extensive medium the gravitational effects of the medium on the body also become important. In this case the passage of the body causes an uneven distribution of the medium due to tidal effects and the clumps of mass so formed will have a direct affect on the body due to their gravitational attractions.

30.5 The Evolution of Planetary Orbits

With the aid of modern computers it is possible to calculate the way that orbits evolve with time. In 2003, I carried out such calculations, modelling the medium by a distribution of point masses orbiting the star. While the term 'disk' is usually used to describe what exists around new stars it is not possible for surrounding gaseous material to take on the form of a true disk of uniform thickness, with a shape like a coin. For equilibrium of the various forces acting, the thickness of the disk must increase with increasing distance from the star and, at any particular distance, the density of the medium must decrease with distance from the mean plane of the disk. Figure 30.3 shows a plan view and a cross section of the disk I used in my model.

In the simulation the medium particles exert gravitational forces on a model orbiting planet and the planet exerts forces on the medium particles. The most common behaviour is that the orbit both rounds off (becomes more circular) and also decays (becomes smaller). One factor that has to be taken into account is the lifetime of disks, due to a decline of the resisting medium with time. This is affected by outward forces due to radiation from the star and also by the stellar wind, a stream of charged particles emitted by the star. When stars are young both these forms of emission can be much greater than when they reach the main sequence. There is also a process of evaporation whereby the material in the resisting medium gradually escapes into space. We have reasonable estimates of the lifetime of the resisting medium from the inferred lifetimes of disks around new stars — from one to a few million years at most (Section 16.3).

There is no way of knowing exactly what the initial orbits were of the protoplanets that eventually gave rise to the Solar System. However, it is possible to show that orbits can round off from very large eccentric orbits to the kind of orbits we see today. The results of one calculation will illustrate this. A protoplanet with the mass of Jupiter starts with an orbit with semi-major axis 2,500 au and

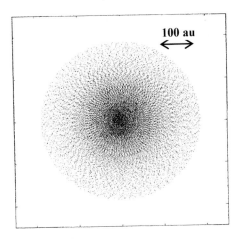

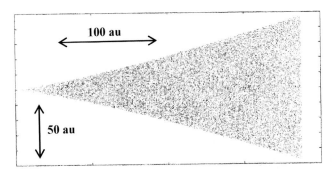

Figure 30.3 The distribution of particles representing the medium in the mean plane of the disk (above) and in a cross section of the disk (below).

eccentricity 0.9. To appreciate how big that orbit is, the aphelion is 160 times as far from the Sun as Neptune, the outermost planet. In the calculation the resisting medium around the Sun is taken as having a mass fifty times that of Jupiter (0.05 $M_☉$) with its greatest concentration close to the star and gradually falling away with increasing distance from the star. The resisting medium also dissipates in such a way that it halves its mass every 1.7 million years — i.e, after 1.7 million years one half of it remains and after 3.4 million years one quarter of the original medium is present. After 3.7 million years the protoplanet has settled down into a perfectly circular orbit with radius 5.3 au, very similar to that of Jupiter. The changes of the

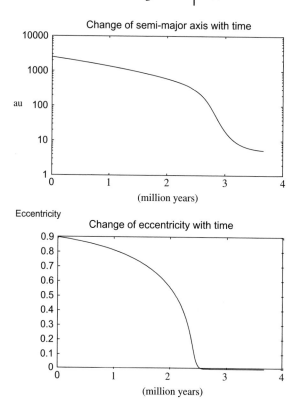

Figure 30.4 Change of semi-major axis and eccentricity with time in a resisting medium.

semi-major axis and eccentricity with time are shown for this calculation in Figure 30.4. The variation of the semi-major axis is so large that the vertical plot is made in logarithmic form. This means that changes by a factor of 10, e.g., from 1,000 au to 100 au, from 100 au to 10 au or from 10 au to 1 au all correspond to the same distance along the axis. The behaviour of the medium during the orbital decay is similar to that displayed in Figure 23.3.

If several planets are produced by a capture-theory process, as must have happened if the Solar System was produced that way, then each planet would have been affected by the resisting medium and ended up in a more-or-less circular orbit much closer to the star (or Sun) than where it was formed.

Observations of planets around other stars show that some of them are in very close orbits. The total decay of the planet depends on the mass of the medium, its distribution and its duration. The higher the density of the medium within which the planet moves, the higher will be its rate of decay and the longer the duration of the medium before it dissipates the greater will be the total decay. In the calculation leading to Figures 30.4 the mass of the medium was taken as fifty times that of Jupiter but some capture-theory simulations give a retained medium with greater mass. That would imply a greater rate of decay of the orbit. On the other hand, observation suggests that some disks decay much more rapidly than the rate taken in the previous section, which would reduce the total effective time for decay to take place. In several of the decay simulations I found final semi-major axes less than 0.1 au, as is observed for some exoplanets. The operation of the mechanism described in Section 23.5 could prevent the planets from actually plunging into the star.

Of great interest are a few observed exoplanets with very eccentric orbits and to understand how these orbits could occur we need to consider in more detail the way in which the forces due to the medium affect the orbit.

30.6 Slowing Down and Speeding Up

The usual assumption of those working on planet migration (Chapter 23) is that the resisting medium is in free orbit around the central star — that is to say that all parts of it are in circular orbits corresponding to that of a planet at the same distance. For a planet in a *circular* orbit this means that local material, if undisturbed, would be moving closely with the same speed as the planet, a little faster inwards and a little slower outwards. The main agencies for orbital change would be the Type I and Type II migration mechanisms, described in Section 23.3, that are mainly influenced by the mass of the planet and the density of the medium.

For planets in an *elliptical* orbit there are considerable relative speeds of the planet with respect to the medium, to the greatest extent at periastron (equivalent position to perihelion for an orbit around a star) and apastron. This gives some, but not significant, viscous resistance and further resistance will be due to the effect of the medium impinging directly on the planet, like wind impinging on the sail of a sailing vessel. However, whatever the relative importance of the various ways for the medium to apply force on the planet, we can make the general observation that the force is always in a direction that will oppose the motion of the planet relative to the medium and that the force is greater for greater relative speed.

Figure 30.5 illustrates the motion of a protoplanet in an elliptical orbit in a freely-rotating medium.

At *periastron* the planet moves faster than the medium and so is slowed down. The effect of this is to change the orbit to one with less energy (smaller semi-major axis) but the same periastron. This makes the orbit smaller and reduces the eccentricity. At *apastron* the planet moves slower than the medium so it is speeded-up. The effect of this is to change the orbit to one with more energy (larger semi-major axis) but the same apastron. This makes the orbit larger but again reduces the eccentricity. Hence, at both extremes the effect is to round off the orbit but there are opposite effects on the energy, hence size, of the orbit. If, as is more common, the density of the medium is larger closer in then the periastron effect will be the stronger and the orbit will round off and decay. Of course, there are resistance forces on the planet at all points on its orbit but the essential features

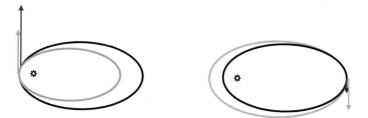

Figure 30.5 Elliptical motion in a freely-orbiting medium. The orbital speeds of the planet at periastron and apastron are marked in red. The medium speed at periastron is shown as green and the effect of the different speeds slows down the planet and modifies its orbit to the green form. Similarly the apastron speed of the medium is shown blue and the effect of the different speeds increases the speed of the planet and modifies the planet orbit to the blue form.

of the orbital modification can be understood just by considering the effects at periastron and apastron.

30.7 Eccentric Orbits

The analysis in the previous section depended on the relative speeds of the planet and the medium at periastron and apastron which, in their turn, assumed that the medium was in free orbit around the star under the sole influence of the stellar mass. It has already been mentioned that new stars go through a very active stage where they are more luminous and have stronger solar winds than when they are on the main-sequence. For example, in 1998 the Italian astronomers, Francesca D'Antona and Italo Mazzitelli, estimated that the early Sun could have been 60 times as luminous as now, with solar winds between ten thousand and one hundred thousand times as strong as now. A strong early stellar wind could have had significant effects on the resisting medium by applying an outward force on it that opposed the gravitational attraction of the star. For the strongest stellar winds, which are present in young stars in their so-called T-Tauri stage, the wind would completely overcome the gravitational attraction of the star and the resisting medium would be driven outwards. Here we are just going to consider the situation where the stellar wind neutralizes some part of the stellar attraction so that the net effect on the medium (but not on the planet which is too large and massive to be greatly affected) is as though the star had a reduced mass. In this case the medium rotates more slowly than if the stellar wind were absent and we shall consider what could happen to an orbit in this case. In Figure 30.6 there is shown the situation where the medium has been heavily slowed down and we show the speeds of the planet and the medium at periastron and apastron.

At periastron the effect is as previously described — the planet is slowed, the orbit decays and it becomes less eccentric. Now, at apastron, the planet is again slowed, the orbit decays and the eccentricity *increases*. The decay is now a consistent feature at both extremes of the orbit, but the changes in eccentricity oppose each other. In the most common situation the density is higher and the resistance force

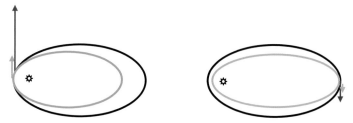

Figure 30.6 Elliptical motion in a strongly slowed-down medium. The orbital speeds of the planet at periastron and apastron are marked in red. The medium speed, reduced by the stellar wind, is shown at periastron as green and the effect of different speeds modifies the planet orbit to green form. At apastron the medium speed, shown in blue, is so reduced by the stellar wind that it is now less than the planet speed. Hence the planet is slowed by the medium and the planet orbit is modified to the blue form.

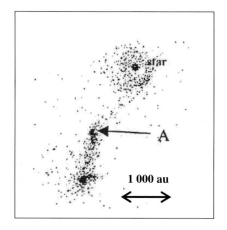

Figure 30.7 A capture-theory simulation showing a strong doughnut-like captured medium.

stronger at periastron so its effect dominates and the orbit is rounded off. However, in many of the capture-theory simulations that have been carried out, the captured material that forms the resisting medium takes up a doughnut-like form (Figure 30.7) so that when the orbit reaches a certain stage in its development the medium density is *higher* at apastron than at periastron and it is the effect there that dominates, so that the eccentricity *increases*. This effect was studied in numerical simulations I carried out and some results are

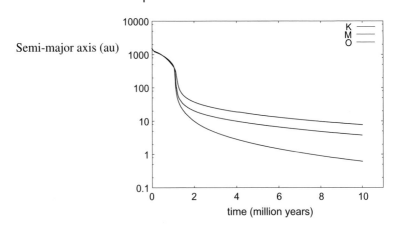

Semi-major axis (au)

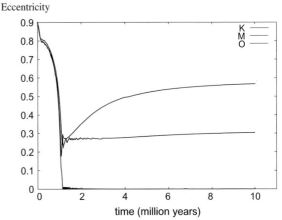

Eccentricity

Figure 30.8 Three simulations of orbital evolution, two of which give eccentric orbits.

illustrated in Figure 30.8. The mass of the resisting medium was 50 M_J but a larger mass, which observations would comfortably allow, could considerably shorten the timescales.

In Figure 30.8, simulation K gave a circular final orbit but the other two end up as ellipses, one with an eccentricity of nearly 0.6 and even higher eccentricities are possible.

30.8 Orbital Periods in Simple Ratios

In the process of the rounding-off and decay of the orbits the proto-planets were not only influenced by the Sun and the resisting medium but also, in some circumstances, by each other. That this might be so is suggested by the fact that the ratios of the orbital periods of pairs of the major planets are very close to the ratios of small integers. For example,

$$\frac{\text{Orbital period of Saturn}}{\text{Orbital period of Jupiter}} = \frac{29.46 \text{ years}}{11.86 \text{ years}} = 2.48 \approx \frac{5}{2}$$

and

$$\frac{\text{Orbital period of Neptune}}{\text{Orbital period of Uranus}} = \frac{164.8 \text{ years}}{84.02 \text{ years}} = 1.96 \approx \frac{2}{1}.$$

In 1996, Mario Melita and I found that there is a mechanism that operates when the orbits of pairs of planets have become circular and are decaying at different rates. When the orbits become *commensurate*, that is that the ratio of their periods equals the ratio of two small integers, then an energy exchange takes place between them. This works in such a way that although the two planetary orbits continue to decay they do so coupled together so that the ratio of their orbital periods remains constant. The effect is a fairly subtle one; while a qualitative theoretical explanation is possible the mechanism is best explored by calculations. It might be asked why it does not operate between Uranus and Saturn (the ratio of periods here is 2.85). Given that the resisting medium has a finite lifetime then it is possible that it simply did not last long enough for a Uranus:Saturn commensurability to become established — although, given time, the ratio may have become 3.0. Alternatively, the comparatively rapid establishment of commensurability between neighbouring pairs of planets may have inhibited the establishment of a further commensurability between one member of each pair.

Chapter 31

How Many Planetary Systems?

So many worlds, so much to do...

Alfred, Lord Tennyson (1809–1892)

31.1 More About Embedded Clusters

Before embarking on an analysis to determine the proportion of stars acquiring planets, it is desirable to describe another unit of length because it is normally used by astronomers to define star number densities (number of stars per unit volume). This unit is the *parsec* (pc), which is 3.09×10^{16} m or 3.26 ly. It is a unit that arises naturally from a method of measuring the distance of close stars by observing them six months apart against the background of more distant stars, observations taken when the Earth is at opposite ends of an orbital diameter. This was the method suggested originally by Kepler (Section 4.4) but which he could not exploit because his measuring instruments were not sufficiently precise. Using this unit of length the star number density in the solar vicinity can be expressed as 0.08 pc^{-3}, that in a normal galactic cluster about 30 pc^{-3}, that in embedded clusters usually in the range $10^3 - 10^4$ pc^{-3} although in the core of the Trapezium Cluster, a rich star-forming region within the Orion Nebula, it is several times 10^4 pc^{-3}.

In Chapter 28 the process was described by which a cloud collapsed, together with the stars in it, to produce the dense embedded stage of a galactic cluster. This dense state lasts for about 5 million years, until supernovae from short-lived massive stars drive out the gas which gravitationally binds the cluster together, at which stage the cloud re-expands. It is thought that dense embedded clusters with present star number densities in the range 10^3–10^4 pc^{-3} could have had precursor states with densities of 10^5 pc^{-3}, something suggested by Bonnell and colleagues in 2001 — although it also seems possible that at least some of the embedded clusters that are now observed may be evolving *towards* the state of greatest density rather than away from it.

31.2 Observational Indications of the Frequency of Planetary Systems

In Chapter 17 three separate ways of detecting exoplanets were described. The first of these, which involves measuring the Doppler shifts in the light coming from the star, favours the detection of more massive planets in close orbits. The virtue of this method is that it gives quantitative results — even if estimates of planetary mass are just of minimum mass. The second method produces infrared images of dusty disks around main-sequence stars considerably younger than the Sun, and implies the possible presence of planets either from gaps in, or from concentrations of, the dust in the disk. The information from this method is suggestive rather than definitive but does indicate the possible presence of planets at distances from stars considerably greater than that of Neptune from the Sun. The third method, that of directly imaging exoplanets, favours the detection of planets at greater distances from stars, where the glare from the star does not overpower the light coming from the planet.

Advances in the Doppler-effect technique have been made by the use of an instrument called HARPS (High Accuracy Radial-velocity Planet Searcher) mounted on European Space Agency telescopes in Chile. Through its use, in April 2009 exoplanets were detected around a nearby star at distance 20.3 ly, Gleise 581, with a mass about $\frac{1}{3}$ $M_\odot$ and designated as a *red dwarf*. There are at least four planets orbiting this star. One of them, Gliese 581d, has an estimated mass of 7.7 $M_\oplus$ and orbits in the so-called *Goldilocks zone* (neither too hot nor too cold) at a distance that could allow liquid water on its surface, giving rise to speculation that it might support some form of life. Another planet, Gliese 581e, with an estimated mass of just 1.9 $M_\oplus$, is more Earth-like in mass, but so close to the star that it is unlikely to possess a biosphere.

The possible detection of a planet with mass even less than that of Gleise 581e has been made by a very different kind of observation. A prediction from Einstein's General Theory of Relativity, which has been confirmed experimentally, is that light is deflected when it passes close to a massive object. This phenomenon gives rise to a phenomenon known as *microlensing,* illustrated in Figure 31.1. Normally, a star is seen, or detected, by light rays that have travelled in a straight line from the surface of the star to the observer, or instrument. However, if a massive body, the microlensing object, is between the star and observer then light that leaves the star over a small angular spread can enter the eye or detector and so the star will appear brighter (Figure 31.1a). If a star with a planetary companion passes between a background star and the Earth then, as it moves across the field of view, the background star will first appear to get brighter, reach a peak when the star is closest to the line of sight, and then decline in brightness (Figure 31.1b). If the planet happens to pass close to the line of sight during this period then there will a sharp rise

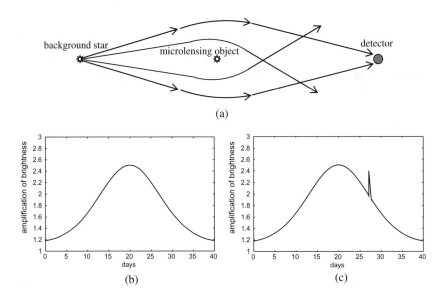

Figure 31.1 An illustration of microlensing. (a) The paths from the background star to detector. (b) The variation of brightness as the microlensing star moves across the field of view. (c) The spike shows the effect of the planet.

in brightness (Figure 31.1c), normally lasting a few hours. The height of the sharp peak enables an estimate to be made of the mass of the planet. In 2009, a planet detected in this way and orbiting a red dwarf was claimed to have a mass of 1.4 $M_\oplus$.

First estimates of the proportion of Sun-like stars with planetary companions were in the range 3–7% but, as techniques have improved with more precise Doppler-effect measurements, direct imaging and microlensing, with the masses of exoplanets that can be detected sharply falling, estimates have steadily risen. Various estimates have been made, some as high as 40%, but it is probably fair to say that any plausible theory should predict the proportion of Sun-like stars with planetary companions as 20% or more. We shall now investigate the frequency with which the capture-theory mechanism occurs within a dense embedded cluster to check whether or not it meets this criterion.

31.3 An Estimate from the Capture Theory

There are a number of parameters that influence the proportion of planetary systems that will form by the capture-theory process. A most important parameter is the density of stars in the embedded cluster for this affects the likelihood that a protostar, or dense region produced by the collision of turbulent elements (henceforth, called simply a 'dense region'), will move close enough to a star to be tidally affected. There is a period of 5 million years or so when the embedded cluster is at, or close to, its maximum density so virtually all planetary formation will occur within that time frame — a conclusion that agrees with the observations of Williams and Cremin (Section 27.4) that star formation in a cluster occurs over a period of a few million years. A model of the way that the stellar number density changes with time, which I have used in various computations, is shown in Figure 31.2; the number density rises to a peak as the cluster collapses and then reduces again when supernovae drive out the binding gaseous component of the cluster, so causing it to expand.

In 2004, at the end of the paper that Stephen Oxley and I wrote describing the basic capture-theory mechanism, we estimated, by a

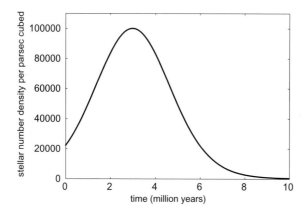

Figure 31.2 A modelled rise and fall of stellar number density in an embedded cluster.

purely analytical process, the proportion of Sun-like stars that should have planets, in which we just considered star-protostar interactions. Observations give estimates of the numbers of stars and protostars to be expected in the embedded cluster at any time so the analysis was quantitative. The proportion found just satisfied the current estimate from observations at that time — 7% — and we concluded that when the contribution from dense regions was added, which probably exceeding the contribution from protostars alone, the criterion would comfortably be satisfied.

Here I describe a different approach using a numerical model. An essential ingredient of the model is that it is based on a great deal of experience in capture-theory simulations. As has already been mentioned, the mechanism is extremely robust and very insensitive in its general behaviour to the initial conditions; unless one uses really outlandish parameters, such as not allowing the protostar closely to approach the star. Under a wide range of conditions the formation of protoplanetary condensations, some of which will be captured, is an inevitable outcome. In the simulation shown in Figure 29.2 the periastron distance of the orbit is less that the radius of the protostar but there is no collision with the star because the protostar is stretched into a filament. Based on experience, it is concluded, as a very *conservative estimate*, that protoplanets will form, with some captured, if the periastron distance is between 0.5 and 1.5 times the radius of the protoplanet.

Although the result of interest is the proportion of Sun-like stars that have planetary companions, what the computational approach was designed to find was the probability that a given *protostar* would interact with a star to give captured protoplanets. The steps in the calculation were as follows:

(i) A star number density in the embedded cluster, n, is randomly chosen in the range $10^4 - 10^5$ pc^{-3}.

(ii) The radius, R, of a spherical region of the cluster is chosen that will accommodate 1,000 stars.

(iii) The 1,000 stars are randomly placed within the sphere.

(iv) The mean speed of stars in an embedded cluster is estimated by the American astronomer, Eric Gaidos, to be in the range 500–2,000 m s^{-1}. For each star a random speed is chosen within that range and the star is given that speed in a random direction.

(v) Each star is assigned a mass, randomly chosen within the range 0.8–2.0 $M_\odot$.

(vi) The protostar is placed at the centre of the sphere with a velocity selected as in (iv).

(vii) The mass of the protostar is selected randomly in the range 0.25–0.75 $M_\odot$.

(viii) Given a temperature 20 K and a mean molecular mass of 4×10^{-27} kg the density and radius of the protostar are found based on the assumption that when formed it is of a Jeans critical mass.

(ix) From the density of the protostar the free-fall time is found from Equation (27.1). The simplifying and somewhat conservative assumption is made that the protostar is only capable of giving a filament for a time, t_i, that is half the free-fall time, during which its radius is constant at 0.9 of its initial radius.

(x) By numerical analysis the motions of the stars and the protostar are followed for a time up to t_i. If during this time the centre of the protostar passes a star at a distance closer than

1.5, but further than 0.5, of its radius then it is deemed to have produced a filament and planets.

(xi) Steps (i) to (x) are repeated 1,000 times and the proportion of protostars giving planets is found.

Figure 31.3 shows the result of running this program in the form of a histogram — in which each block gives the number of closest passages in ranges of one half of the radius of the protostar. There are 92 passages between 0.5 and 1.5 of a protostar radius so that 9.2% of the protostars end up as the producers of planets with the remainder collapsing to produce stars.

This analysis has worked out the probability that a protostar will *give* planets although what we wish to know is the proportion of stars that *capture* planets. From the statistics of stellar masses it is found that there are five times as many protostars with masses in the range 0.25–0.75 $M_\odot$ as there are stars in the mass range 0.8–2.0 $M_\odot$. Assuming that no star experiences more than one capture-theory event this means that the proportion of stars expected to have planets is $0.092 \times 5 = 0.46$, i.e., just under one-half.

The model contains a number of assumptions and approximations. The distinction between protostars and stars on the basis of mass is not really valid; the simulation conditions are based on the

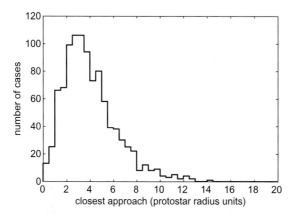

Figure 31.3 The histogram of closest approach distances for 1,000 protostars.

observation that stars tend to decrease in mass as star formation progresses, so that an early-formed star will have a tendency to meet later-produced protostars of lower mass. However, there is no reason to believe that the approximations in the model have unduly exaggerated the estimated proportion of planetary systems and there is plenty of slack between the results found from this numerical study and what observations indicate thus far. Once again we have not included interactions between stars and dense regions that would contribute many additional planetary systems. If there is a fault in the capture-theory model it is that it produces an embarrassingly large proportion of planetary systems although, as we shall see, they will not all survive.

31.4 The Disruption of Planetary Systems

We have already mentioned that the proportion of stars with planetary systems will depend critically on the number density of stars in the embedded cluster — the higher the number density the greater the proportion there will be of stars with planetary companions. We have also noted that when protoplanets are first formed they are on extended orbits that take them to large distances from their parent stars, particularly at apastron where they spend a substantial proportion of each orbital period. In such environments they are easily perturbed by near-passing stars to the extent of becoming unbound from their parent stars. The high number density of stars that was so helpful in producing planetary systems has now becomes extremely unhelpful in retaining them! If the planet were permanently on its extended orbit, with the condition of highest star number density in the cluster, then the probability of its survival in the long term as a planetary companion would be zero. On the other hand, if it were permanently in a condition when the orbit has evolved to its final state and the star number density has fallen to a low value then it is virtually certain that it would exist as a planetary companion indefinitely. So, it is all a question of timing. The true situation lies between these two extremes and the probability of orbital disruption depends on the rate at which the orbit changes and the variation of the star number density with time, which governs the probabilities of stars approaching each other at various distances. In 2004, I carried out an analysis to estimate the probability that a planet produced by the capture-theory mechanism would survive. It involved carrying out a large number of individual calculations in which a planet in an orbit with semi-major axis, *a*, and eccentricity. *e*, corresponding to a particular stage of the orbital evolution of a particular planet, is perturbed by a star, whose mass must be defined, passing on an orbit bringing it to closest distance, *D*, from the parent star. Figure 31.4 shows the situation with the two stars at their closest approach. However, the plane of the planet's orbit will, in general, differ from that of the perturbing star's orbit so this is another variable to take into account, together with the orientation of the orbit, defined by the direction of its line of apses, the line joining periastron to apastron. Again, what happens to the planet depends on where it is in its orbit at the time of closest approach of the two stars — yet another variable (for some positions the planet will survive while for others it will be lost). Finally, the time of formation of the planetary system is an important parameter; if it were produced when the stellar density was highest then there would be the greatest likelihood of disruption.

Once the parameters have been defined then individual calculations are straightforward and quickly carried out. There are three

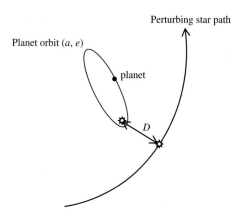

Figure 31.4   The passing star at closest position to the parent star.

possible outcomes. The first is that the planet is removed from its orbit to become a free-floating planet. The second is that the planet is retained in orbit but the semi-major axis is increased so that the orbit has *softened* (Section 28.3). The third is that the semi-major axis is reduced and the orbit has *hardened*.

It will be appreciated that turning the results of many individual calculations into an average proportion of planets that will be lost is quite a complex process. The conclusion reached was that the proportion of planets retained was 0.48 but, taking into account reasonable variations in the assumptions being made, it could fairly reliably be put in the range one-third to two-thirds. For the planets that were retained, about one-third of the orbits hardened and the remaining two-thirds were softened. Hardening and softening had no substantial effect on the proportion of stars with planets; the main effect was to modify the time it took for the planetary orbits to evolve.

It must not be thought that if, say, two-thirds of planets are lost then two-thirds of planetary systems are lost. Consider a star that has acquired four planets. If the probability of losing each planet is two-thirds and the probabilities for the individual planets are independent, then the probability that *all four* are lost is $(\frac{2}{3})^4 \approx 0.20$, i.e., there is an 80% chance that the star retains *at least* one planet. Capture-theory simulations suggest an average of about four planets captured so the reduction in the predicted proportion of stars with planets, estimated as 46% in Section 31.2, would only be reduced to just below 37%.

Inevitably, in modelling complex situations — such as the formation of planetary systems by the capture-theory process under a variety of conditions and their subsequent disruption by stellar perturbations, also under a variety of conditions — it is impossible to find precise estimates. Consequently, the estimates given in this chapter must be considered as ballpark figures. While the 0.46 estimate that the initial proportion of stars acquiring planets cannot be regarded as completely reliable what *is* reliable is the statement that planetary formation by the capture-theory process is not a rare event and that, on the contrary, it happens easily and often and that there is a high probability that a planetary system, once formed, will survive, at least in part.

Chapter 32

Starting a Family

All happy families resemble one another, but each unhappy family is unhappy in its own way.

Leo Tolstoy (1828–1910), *Anna Karenina* (1878)

32.1 Satellites and Angular Momentum

All the major planets of the Solar System possess extensive families of satellites, a fact that suggests that satellite formation may be a normal concomitant of major planet formation, including that of exoplanets.

When Galileo first observed the large satellites of Jupiter, now known as the Galilean satellites, with his telescope he saw them as related to Jupiter in the same way as the planets are to the Sun and this reinforced his belief in the Copernican theory. This idea, that satellites are related to planets as planets are to the Sun, has often been expressed. In 1919, Jeans stated that any theory that proposed a mechanism for producing satellites that differed from that for producing planets would be "condemned by its own artificiality". Again, in 1978, the Swedish astrophysicist and Nobel Laureate, Hannes Alfvén, stated that, "We should not try to make a theory of the origin of planets around the Sun but a *general theory of the formation of secondary bodies around a central body*. This theory should be applicable both to the formation of satellites and the formation of planets." When three eminent scientists, widely separated in time, all express the same view then it tends to be generally accepted. Indeed, the SNT proposes that satellites would form in disks around planets much as planets form in disks around stars, which fits in with the generally accepted view. We now critically examine this belief.

We have already seen that Laplace's nebula theory and Jeans' tidal theory were both rejected on grounds relating to angular momentum, although for different reasons. Laplace nebula model could not explain the small angular momentum of the spinning Sun while Jeans' tidal theory did not give enough angular momentum to the orbiting planets. It is instructive to look at the relationship between the angular momentum associated with the *primary body* — the Sun or a planet — and the *secondary body* — respectively, a planet or a satellite. We characterize the angular momentum of the primary body by the intrinsic angular momentum (angular momentum per unit mass) for material at its equator and the angular momentum of the secondary body as the intrinsic angular momentum of the body in its orbit. Table 32.1 shows the ratio of these intrinsic angular momenta for various primary-secondary combinations.

It is clear that, in this respect, there is a significant difference in the two types of system. The smaller values of S for planet-satellite systems make it more plausible that they could be derived from the same body of material whereas, equally clearly, if the Sun and planets are to come from the same body of material then some significant extra process, or processes, must come into play.

The basic SNT mechanism produces both the primary and secondary bodies from the same body of material so here we investigate whether the same mechanism can explain regular satellite formation, without the timing difficulties it encounters in producing planetary systems.

Table 32.1 The ratio, S, of the intrinsic angular momentum of the secondary orbit to that of the spin of the central body at its equator

Central body	Secondary body	Ratio S
Sun	Jupiter	7800
Sun	Neptune	18700
Jupiter	Io	8
Jupiter	Callisto	17
Saturn	Titan	11
Uranus	Oberon	21

32.2 Dust Settling in the Disk

Figure 29.3 shows that as a protoplanet formed by the CT collapsed, it formed a dense core, destined to become the planet, and a surrounding disk. The figure only shows the initial stages of this process; the core is collapsing towards planetary dimensions and the surrounding disk is also in a state of collapse. Following the SNT process for producing planets the first stage for satellite formation must be the settling of the dust in the disk to form a dust carpet.

In the work by Weidenschilling and his colleagues (Section 21.2) the size of the dust grains were usually taken as having diameters of order 1 μm and the time deduced for the settling of such grains was close to the limit of the lifetime of disks. To overcome this problem the idea was proposed that the dust grains adhered to each other to form aggregates that would settle more quickly.

The question of dust grain size is worth looking at in more detail in the light of current knowledge. Dust grains reveal themselves to observers on Earth in two ways — by blocking out light and by reddening transmitted light due to the preferential scattering of light at the blue end of the spectrum. For a given mass of dust both these effects are more effective for smaller grains. However, from observations by the NASA/ESA Ulysses spacecraft, which collected interstellar dust, it has been concluded that there is a distribution in the sizes of dust grains, with sizes up to 5 μm or more, with the numbers of grains of different masses, m, varying as $m^{-1.63}$. It can be shown that the consequence of this distribution is that, while absorption and scattering is dominated by small grains, it is the grains at the upper end of the size range, above 2 μm, that contain most of the mass. Thus, in forming a dust carpet it is the larger grains, which settle more quickly, that are of primary concern.

The size and distribution of the material in a disk is of great importance in determining the form of its development. In 2005 the Portuguese astronomers, S. Vicento and J. Alvez, found that 60% of the circumstellar disks for newly-forming stars in the Trapezium Cluster, situated in the Orion Nebula, had radii less than 50 au, with some radii up to a few hundred au. Hence, it is likely that the disk round the early Sun had a radius a few times that of the eventual radius of the Sun's planetary system. The radius of the orbit of Callisto, the outermost Galilean satellite is 1.88×10^9 m so, scaling from the circumstellar disk, we take the radius of a circumplanetary disk as of order 10^{10} m. Actually, the radius of a disk is impossible to define precisely as the disk density will gradually reduce with increasing distance from the planet or star. A circumstellar disk will, at some distance from the star, become undetectable from Earth and this is the distance that would be estimated as the radius of the disk.

The theory developed by Weidenschilling can be used to calculate the settling times for dust at various distances from a planet. The example taken here has the following parameters:

Mass of planet = 2.0×10^{27} kg (approximately the mass of Jupiter)
Mass of disk = 2.0×10^{27} kg
Mean molecular mass of gaseous material = 4×10^{-27} kg
Temperature of disk material = 20 K
Density of dust grains = 3×10^3 kg m^{-3}
Ratio of principle specific heats of the gas = 5/3

The areal density of the disk was taken as falling off with distance from the planet so that it fell by a factor of e^1 for every 2×10^9 m, so that at a distance of 10^{10} m from the planet the density will have declined by a factor of about 150. The settling time to the mean plane was found for particles of radius 2, 3, 4 and 5 μm for distances between 3×10^9 m and 6×10^9 m from the planet. Due to the flared nature of the disk (Figure 30.3) dust particles further out have a greater distance to fall but the results, shown in Figure 32.1, show that they reach the mean plane faster, because the larger distance they have to fall is more than compensated for by the lower density of the gas they fall through. A mechanism we have not considered, which would reduce times even more, is that because larger grains

---

[1]The exponential e (=2.7182818) occurs in a natural way in mathematics and is used to describe exponential decline. The function e^{-x} reduces in value by a factor $1/e$ for each increase of x by 1.

fall faster than smaller ones then, as they fall, they sweep up smaller grains, grow in size and hence fall even faster. If this factor were taken into account the relationship of settlement time to distance seen in Figure 32.1 is reversed and the settlement time is faster in the *inner* part of the disk. This is because the disk is denser further in and particles grow more rapidly so that by the time they reach the mean plane they have grown to a radius of more than 50 μm.

Discounting grain growth, for a conceivable disk lifetime there is no carpet formation less than about 3.5×10^9 m from the planet and for a reasonable disk lifetime of one million years carpet formation will only be complete beyond about 5×10^9 m from the planet. Another criterion to be satisfied is that there must be sufficient material to form satellites — for example, a mass of 4×10^{23} kg to form the Galilean satellites. For the disk we are considering the total mass between a distance of 5×10^9 m from the planet and the nominal edge of the disk at a distance of 10^{10} m is 4.94×10^{26} kg and if just 0.5% of that settles in the dust disk the mass of the carpet would be 2.5×10^{24} kg — ample to form the Galilean satellites.

In the light of the above information we will consider how the Galilean satellites may have formed, assuming a limited disk lifetime so that only material beyond a distance of 5×10^9 m is involved, beyond which distance a dust carpet will form.

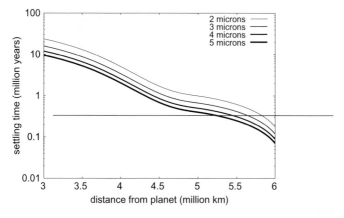

Figure 32.1 The settling times for dust particles at different distances from the planet.

32.3 The Formation of Satellitesimals

The next stage in the process of forming satellites is for gravitational instability in the disks to form solid clumps that, following the SNT terminology, we call *satellitesimals*. From the standard Ward and Goldreich theory (Section 21.3) the masses of the satellitesimals as a function from distance from the planet are shown in Figure 32.2.

Comparing satellitesimal masses with masses of Jupiter's satellites, the largest mass of a satellitesimal, at 5 million km is about 1.5% of the mass of Io. The next stage of satellite formation involves the accumulation of satellitesimals, for which we have the theory developed by Safronov (Section 21.4). This gave timescale problems for the SNT. We shall have to see if it is equally troublesome for satellite formation.

32.4 Satellite Formation

Safronov's Equation (21.2) for the time required for planetesimals to accumulate to form a terrestrial planet, or the core of a major planet, can also be applied to the time for satellitesimals to produce a satellite. To get a ballpark figure for the time of formation of a Galilean satellite we take an average satellite of final radius 2,000 km, and density 2,500 kg m^{-3} with the disk characteristics given in Section 32.2,

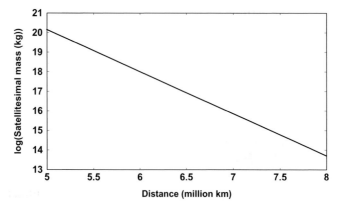

Figure 32.2   Satellitesimal masses at various distances from the planet.

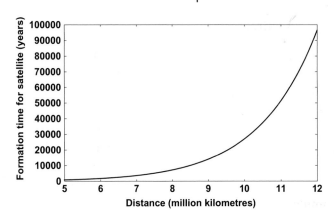

Figure 32.3   Formation times of satellites from satellitesimals at different distances from planet.

assuming that 1% of the disk material is dust. The time of formation for different distances varying between 5 and 12 million kilometres from Jupiter are shown in Figure 32.3.

The times of formation are short, especially when compared with the long formation times for planets according to the basic SNT. This is because all the factors in Equation (21.2) differ from those in the planet-formation case to shorten the time. The size and densities of the bodies are both somewhat smaller, the periods of their orbits considerably smaller and the average areal density of satellitesimal material in the satellite-forming regions is much larger.

The timescale estimates we have made here are based on parameters that have a great deal of uncertainty but they would have to be grossly changed to make the formation times of satellites a problem in relation to the probable lifetime of the disk.

32.5 Migration Processes

The distances at which we have been considering satellite formation, up to 12 million km from the planet, are much larger than the orbital radii of the Galilean satellites — from 0.422 million km for Io up to 1.880 million km for Callisto. However, for small satellitesimals in the comparatively dense circumplanetary disk, Type I migration would operate very powerfully and the satellitesimals would begin migration inwards as soon as they formed. If we consider the formation of Io, the original satellitesimals to form it might originate at a distance of, say, 6 million km but during the period when the satellite was growing by the addition of more satellitesimals it would also be drifting inwards. In Section 30.8, it was explained how the phenomenon of orbital commensurabilities arose for the planet pairs Jupiter-Saturn and Uranus-Neptune. In the dense medium in which the orbits of the Galilean satellites were evolving the same process would occur very strongly giving the remarkable commensurabilities linking Io, Europa and Ganymede (see Section 6.2).

The compositions of the Galilean satellites, with no ices on Io, a thin layer of water ice on Europa and a considerable ice content for both Ganymede and Callisto, suggest that Io was formed in, or entered into, a temperature environment in which ices could not exist and that the temperature fell with distance from Jupiter allowing a greater amount of ices to be present. At present Jupiter emits about two-and-a-half times as much energy as it receives from the Sun and the probable explanation of this is that Jupiter is slowly contracting and releasing gravitational potential energy. Shrinkage of the radius by 5 centimetres per century would provide the necessary energy. When Jupiter was first produced it would have taken a considerable time to settle down to close to its present state and shrinkage of a few hundred metres per year would have given it a high surface temperature and a high luminosity. In 1982, Nicholas Schofield and I studied the collapse of Jupiter and concluded that for some time after its formation it would have had a luminosity of 10^{23} W. This would give a temperature environment that would have driven all ices out of Io, allowed retention of some water on Europa and allowed ices to remain a part of the other two Galilean satellites.

32.6 Irregular Satellites

Many satellites in the Solar System are much smaller than the Galilean satellites but nevertheless have the characteristics of regular satellites, orbiting in more-or-less circular orbits very close to the equatorial plane of the planet. There are notable exceptions, including the Moon, which will be considered further in Chapter 36, and Neptune's retrograde satellite Triton, to be dealt with in Chapter 38. The majority of irregular satellites are almost certainly captured bodies — for example, Saturn's Phoebe (Table 6.2), which is in a retrograde orbit. Although capture is often cited as the source of most irregular satellites, it is a process requiring some special circumstances. If a small body — a potential irregular satellite — were to approach an isolated planet then, unless there was some agency to remove energy from the two-body system, capture could not take place. This agency could be the presence of a third body, which increases its energy while removing it from the relative motion of the two bodies. Other possible agencies are either tidal action or a collision, which would convert energy of motion into other forms of energy, such as heat. Here, as an example, we shall consider an explanation for the two outer sets of Jupiter's satellites that are listed in Table 6.1.

The two groups of small satellites are at considerable distances from Jupiter. The members of the inner group, orbiting at distances between 11 and 12 million km, have quite eccentric orbits with inclinations between $25°$ and $29°$ while members of the outer group, orbiting between 20 and 24 million km from Jupiter, have retrograde orbits of even higher eccentricity. They are the result of some capture process and it is significant that the apojove[2] distance, 14.18 million km, of Elara, a member of the inner group, is just greater than the perijove distance, 13.75 million km, of Pasphaë, a member of the outer group. Figure 32.4 illustrates the result of a proposed collision between two large asteroids, A and B, both originally in direct orbits around the Sun, in the vicinity of Jupiter. After the collision there are two sets of fragments, each following a path somewhat deviating from the original path although in the same general direction, one set in direct orbits and the other in retrograde orbits around Jupiter.

---

[2] Apojove and perijove refer to the greatest and smallest distances of a satellite from Jupiter.

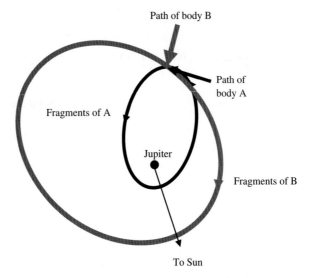

Figure 32.4 A collision of two bodies near Jupiter giving rise to two outer families of satellites. All motions are shown relative to Jupiter.

Of course, any explanation of these irregular satellites and others must be speculative but what is clear is that collisions can explain the presence of most, if not all, irregular satellites.

Chapter 33

Tilting — But not at Windmills

For sideways would she lean, and sing a faery's song.

John Keats (1795–1821)

33.1 The Leaning Sun and Retrograde Exoplanets

An interesting feature of the Sun is that its spin axis is inclined at an angle of 7°. "Inclined to what?", you may ask, and the answer is to the normal (perpendicular) to the plane of the Earth's orbit, the *ecliptic*, the usual reference plane for defining the inclinations of planetary orbits. This arrangement is illustrated in Figure 33.1.

The Sun's spin axis has to point in some direction so it might be wondered why it is that the 7° tilt is thought to be of interest. Actually, it is both interesting and important. For some theories, i.e., the Solar Nebula Theory (Chapter 21) the angle would be expected to be zero and although the actual angle is fairly small it is large enough to require an explanation — for example, in terms of a body passing by the Solar System and disturbing the plane of the planetary orbits. Again, the small angle is unlikely to be due to pure chance; the probability that two random vectors will be 7° or less apart is about 1 in 270.

However, while the 7° tilt may be uncomfortably large for the Solar Nebula Theory it is also uncomfortably small for the Capture Theory (Chapter 29). The plane of the planetary orbits is defined by the plane of the Sun-protostar orbit, which has no systematic relationship to the spin axis of the Sun.

Although the relationship of the Sun's spin axis to the ecliptic is usually described as a tilt of the spin axis, it might equally be described in terms of the inclination of the planetary orbits so that, for example, the Earth's orbit has an inclination of 7°. This way of

297

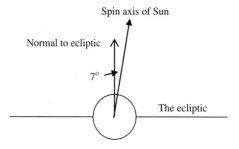

Figure 33.1 The Sun's spin axis in relationship to the ecliptic.

describing the situation has the advantage that the orbital inclinations of some exoplanets have been found, defined in relation to the spin axis of the parent star, and then we can consider the Solar System as just another member of the family of all planetary systems and not regard it as a special case.

In Section 17.7 we reported that of 27 exoplanet orbital inclinations that have been measured six were retrograde; if we now include our Solar System in the sample it is six out of 28, or 21%. We now address the question of how to explain this state of affairs. Clearly it presents a problem for the SNT that would certainly initially produce planets in prograde orbits — and, what is more, with zero, or at most small, inclinations. To solve this problem for the SNT various mechanisms have been proposed. Two of them involve changing the plane of the planet's orbit either by interaction with another planet or by the action of an external star by what is known as the Kozai mechanism. A third idea, proposed by T.M. Rogers and his colleagues, suggests that a process occurs within a star that can cause the surface material of the star to spin about a different axis from that of the bulk of the star — which makes the exoplanet orbit apparently, but not really, have a high inclination if it is taken with respect to the spin of the whole star.

To help our discussion of this topic we have to expand our knowledge of the nature of angular momentum. It is a *vector quantity*, which means that it not only has a magnitude but can also be associated with a direction. For the case of a body like the Sun, or a general star, the direction is defined by the spin axis. For an orbiting planet the direction of the angular-momentum vector is perpendicular to its orbital plane. If we represent the angular-momentum vector by an arrow then the convention is that if we look down the axis so that the arrow points towards us and the rotation is counter-clockwise then it is prograde; if clockwise the rotation is retrograde.

Now we consider the question of orbital inclination in relation to the CT. In Figure 33.2 we show the angular momentum vector of a star, represented in direction by arrowed line A, and that of an exoplanet, defined by the plane of the star-protostar orbit and represented in direction by arrowed line B. In Figure 32.2a the exoplanet orbit is prograde ($i < 90°$) and in Figure 33.2b it is retrograde ($i > 90°$).

In Chapter 30 we dealt with the way that protostar material, captured in the form of a circumstellar disk, acted as a resisting medium

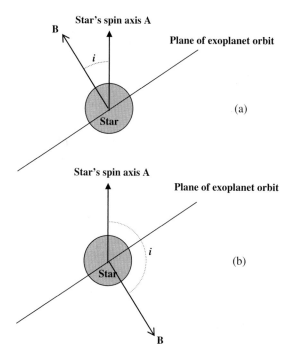

Figure 33.2   Possible arrangements of the plane of an exoplanet orbit and a star's spin axis in a capture-theory interaction giving (a) a prograde exoplanet orbit and (b) a retrograde exoplanet orbit.

to modify the initial protoplanet orbits, in general giving round-off and decay. Figure 30.7 shows captured material for a case where the captured material had a doughnut form, but more frequently the medium fell off gradually from a peak close to the star. What the resolution of the simulation could not reveal is that some material actually impinged on the star and was absorbed by it. The amount that was absorbed depended on the parameters of the star-protostar interaction — it could be very little or a great deal — and this would have a strong influence on the final inclination of the exoplanet orbit.

33.2 Moving the Star's Spin Axis

The angular momentum associated with the absorbed protostart material combines with that of the star and changes the angular momentum of the star's spin both in magnitude and direction. The way that the angular momentum of the added material will combine with that of the star to give a resultant angular momentum, both in magnitude and direction, is illustrated in Figure 33.3. The length of the arrows gives the magnitude of the angular momentum and their direction indicates the associated spin axis. What is seen in Figure 33.3 is that the star's spin axis is pulled towards the axis associated with the star-protostar orbit. The extent to which this happens depends on the angle between the original angula momenta and the relative magnitudes of the two components that give the resultant angular momentum.

An important aspect of this mechanism for reducing the tilt of the solar spin axis is that comparatively little protostar material needs to be absorbed by the star. As an example, the present angular momentum of the Sun is equivalent to one-third of a Jupiter mass in orbit at its equator and this will be approximately the case for many Sun-like stars. Given that the expected mass of a resisting medium is expected to be several tens of Jupiter masses it is clear that a very small proportion of the material coming from the protostar needs to be absorbed by a star to give a large effect on the stellar spin axis.

Now we consider by how much the direction of the star's spin axis can be changed. We take the magnitude of the angular momentum

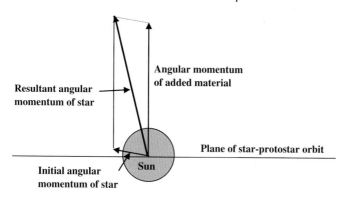

Figure 33.3 The change of direction of a star's spin axis due to the addition of protostar material.

of the added material as η times that of the star. If η is almost zero then, regardless of its direction, it will have virtually no effect on the direction of the spin axis of the star. In that case the likelihood of a retrograde orbit would be 50%, which is clearly not what is observed. However, if η is extremely large then the final spin axis of the star will be close to the normal to the exoplanet orbital plane and, with respect to the final spin axis of the star, the inclination of the exoplanet orbit will be small; in this case there would be no retrograde exoplanet orbits.

For any value of η the result of adding the absorbed material in terms of changing the inclination of the exoplanet orbit will depend on the relationship between the original direction of the stellar spin axis and the plane of the exoplanet orbit. If, for example, the stellar spin axis happened to be orthogonal to the plane of the orbit then the direction of the stellar spin axis will be unchanged regardless of the value of η. For any particular value of η and relative configuration of the spin axis and orbital plane the final inclination of the exoplanet orbit can be found. Taking the relative inclination as a random quantity it is possible to find a probability distribution of the values of the final orbital inclination, *i* for any given η. It turns out that the form of the distribution is very different for the two ranges η > 1 and η < 1 with η = 1 having some characteristics linking it with each range.

η > 1

Theory shows that in this case the final exoplanet orbit cannot be retrograde, so we must have $i \leq 90°$, and the maximum possible inclination, i_{max} falls as η increases. Figure 33.4 gives computational results for $\eta = 1, 2, 4, 6, 8$ and 10. It can be seen that for $\eta = 1$ the value of i_{max} is 90°, the maximum value allowed for this range of η. The likelihood of getting different inclinations is expressed by $P_\eta(i)$, the probability of having inclination i per angular degree range.

Various characteristics of the distributions $P_\eta(i)$ were found and are given in Table 33.1. They are:

- i_{max}, as shown in the figure and which can be computationally determined,
- i_η, the mean value of i and
- P_η (7) the value of $P_\eta(i)$ for $i = 7°$, which is the mean inclination of solar-system planets relative to the Sun's spin axis.

An unexpected result found from the computation was that

$$\eta \times \overline{i_\eta} = \text{constant} = 45°. \qquad (33.1)$$

Table 33.1 Characteristics of $P_\eta(i)$ for $\eta \geq 1$

η	$i_{max}(°)$	$\overline{i_\eta}(°)$	$P_\eta(7)$
1.0	90.00	45.00	0.00466
2.0	30.00	22.50	0.01110
3.0	19.47	15.00	0.02280
4.0	14.48	11.25	0.04114
5.0	11.54	9.00	0.06920
6.0	9.59	7.50	0.11484
7.0	8.21	6.43	0.20553
8.0	7.18	5.62	0.77027
9.0	6.38	5.00	0.00000
10.0	5.74	4.50	0.00000

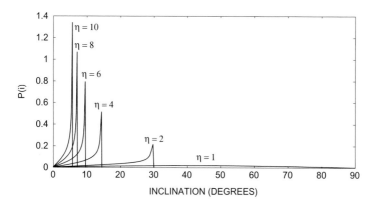

Figure 33.4 The probability distribution, $P_\eta(i)$, for $\eta \geq 1$.

$\eta \ll 1$

The distributions for $\eta = 0.0, 0.2, 0.4, 0.6$ and 0.8 together with that for $\eta = 1.0$ are shown in Figure 33.5. All values of i between $0°$ and $180°$ are possible for $\eta < 1$. It will be seen that with increasing η the curve collapses towards the i-axis beyond $i = 90°$ and for $\eta = 1$ the curve becomes zero beyond that value.

Characteristics of the distributions for $\eta \leq 1$ are given in Table 33.2 and include p_{retro}, the proportion of orbits that are retrograde, given by the total probability that $i > 90°$. Again there is a simple relationship between $\bar{i_\eta}$ and η but of a different kind from that found

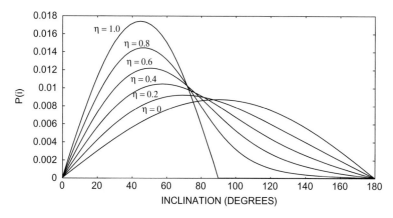

Figure 33.5 The probability distribution, $P_\eta(i)$ for $\eta \leq 1$.

Table 33.2 Characteristics of $P_\eta(i)$ for $\eta \le 1$

η	$\overline{i_\eta}(°)$	$P_\eta(7)$	p_{retro}
0.0	90.00	0.00106	0.500
0.1	85.50	0.00128	0.450
0.2	81.00	0.00152	0.400
0.3	76.50	0.00179	0.350
0.4	71.00	0.00207	0.300
0.5	67.50	0.00237	0.251
0.6	63.00	0.11270	0.201
0.7	56.50	0.00305	0.151
0.8	54.00	0.00341	0.101
0.9	49.50	0.00380	0.051
1.0	45.00	0.00421	0.000

for $\eta > 1$; now $\overline{i_\eta}$ varies linearly between $\eta = 0$ and $\eta = 1$. Another relationship revealed by the calculation is the linear variation of p_{retro} between $\eta = 0$ and $\eta = 1$, a result that can be confirmed by mathematical analysis.

The single value of η that would give the 21 percent of observed retrograde orbits is 0.58, but, of course, for the different planet-protostar interactions that gave the sample of 28 orbital inclinations we are considering, there would be a variety of values of η, some giving a greater probability of retrograde orbits and some a lesser probability.

33.3 The Inclinations of Solar-System Planetary Orbits

From analysis it is found that an inclination as large as $7°$ can only occur for $\eta < 8.2$, which is consistent with the results given in Table 33.1. The variation of $P_\eta(7)$, the probability per degree at an inclination value of $7°$, with η is shown in Figure 33.6. The value of $P_\eta(7)$ increases up to $\eta = 8.2$, beyond which the value of i_{max} falls below $7°$. The curve indicates that it is likely that the value of η that gave the inclinations of solar-system planets was considerably greater than unity, although lower values cannot definitely be ruled out.

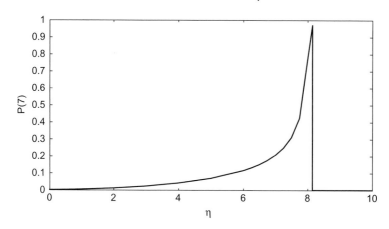

Figure 33.6 The variation of $P_\eta(7)$ with η.

The absorption of some protoplanet material by the star, which can be either little or great, both explains why we have many retrograde orbits, but less than 50%, and why we can also have small inclinations, as is the case for the Solar System.

33.4 A Child's Top and Evolving Planetary Orbits

There was a time, alas no more, when most children played with whips and tops rather than with electronic gadgets. By repeatedly whipping the top it could be kept in a rapidly-spinning state with the spin axis in a vertical position. However, if the spin rate slowed down another form of behaviour would ensue. The axis of the top would now be inclined to the vertical and that axis would undergo *precession* round the vertical direction. This is illustrated in Figure 33.7. The reason for the precession of the spin axis is that the combination of the force due to gravity acting downwards, plus the reaction force acting upwards constituted a torque in a direction which tried to pull the spin axis downwards; the laws of mechanics then operate to give precession.

We now relate the behaviour of the top to the evolution of the elliptical orbit of a planet in the presence of a resisting medium, with forces operating as seen in Figure 33.8. The corresponding motion

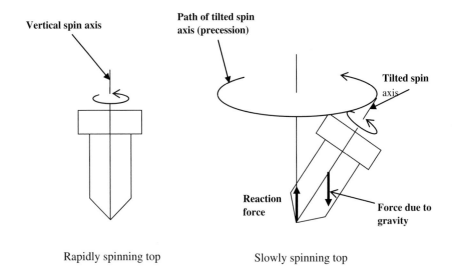

Figure 33.7 The precession of the spin axis of a slowly spinning top with a non-vertical axis.

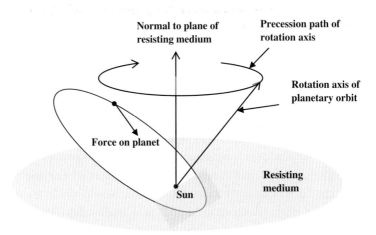

Figure 33.8 The non-central force due to the mass of the resisting medium gives precession of the planetary orbit.

to the spin of the top is the orbit of the planet. As the resisting material has mass and is in the form of a disk, the force it exerts on the planet points away from the point about which the orbit takes place — the Sun. This force tries to rotate the orbital axis and the spin axis undergoes precession about the normal to the resisting medium.

This kind of motion has been confirmed by detailed computer-based calculations.

In the figure showing precession, the angle made by the orbit with the resisting medium is greatly exaggerated for clarity of presentation. In the capture-theory model the initial orbits of the protoplanets would not be exactly in the plane of the protostar orbit because the spin of the protostar, which could be about an axis in any direction, would throw them slightly out of the plane. However, the angle between the orbit and the plane could be at most about 5°.

Calculation shows that the precession of a planetary orbit would be more complicated than that of a top, with other kinds of motion taking place but the important part of the precession motion is as described. From above, projected onto the plane of the resisting medium, the motion would be seen as shown in Figure 33.9.

The periods for these precessions are of the order of a hundred thousand years so, since the planetary orbits take a few million years to settle down to their final state, this means that there will be a several complete precession orbits during the period of complete orbital evolution.

33.5 Deuterium in Major Planets

We are in the process of explaining the tilts of planetary spin axes, for which the preceding description of the precession of the evolving

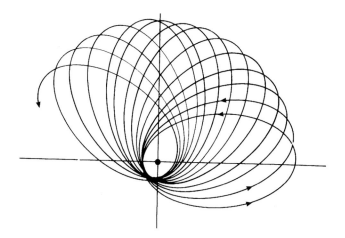

Figure 33.9 A projected view of the precession of a planetary orbit.

planetary orbits is directly relevant. Now, we may seem to be going off at a tangent discussing something completely different. However, I can assure you that it is relevant and, furthermore, this topic will be of great importance in a later event to be described in Chapter 35.

The first and, indeed, only atoms to be formed in the Big Bang that created the Universe were the lightest ones — hydrogen, deuterium, helium and a small amount of lithium. These are shown in schematic form in Figure 33.10. Deuterium is an *isotope* of hydrogen, meaning that chemically it behaves like hydrogen because, like hydrogen it has a single proton in its nucleus and it is the number of protons that determine the chemical properties of atoms. The difference is that deuterium has an extra neutron in its nucleus, giving it twice the mass of hydrogen. Deuterium is important enough to have its own chemical symbol, D, so while water is usually described as H_2O another form of water is HDO and D_2O, known as *heavy water*, is used in the nuclear power industry.

In the Universe as a whole the ratio of the number of deuterium to hydrogen atoms is $D/H = 2 \times 10^{-5}$ but there are wide variations displayed for various solar-system bodies as shown in Table 33.3. The outstandingly high ratio in the table is that for Venus, which is due to a special evolutionary factor for that body, which also led to it being very arid. As it is quite close to the Sun it had a considerable amount of water vapour in its atmosphere that was dissociated by solar radiation by the reaction,

$$\overset{\text{radiation}}{H_2O \quad \rightarrow \quad OH + H}$$

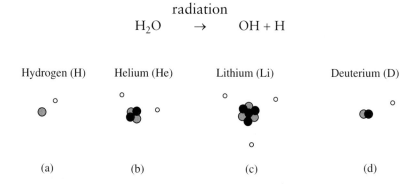

Hydrogen (H)   Helium (He)   Lithium (Li)   Deuterium (D)

(a)   (b)   (c)   (d)

Figure 33.10 The dominant material content of the early Universe (a) hydrogen, (b) helium, (c) lithium and (d) deuterium. Grey circle = proton, black circle = neutron, white circle = electron.

Table 33.3 Deuterium/
Hydrogen ratios for the
Universe and various
solar-system bodies

Body	D/H
Universe	2×10^{-5}
Jupiter	2.3×10^{-5}
Saturn	2.5×10^{-5}
Uranus	5.5×10^{-5}
Neptune	6.5×10^{-5}
Earth	1.5×10^{-4}
Venus	1.6×10^{-2}
Comets	3×10^{-4}
Meteorites	up to 10^{-3}

or, if the water contained deuterium,

$$HDO \quad \xrightarrow{\text{radiation}} \quad OH + D.$$

Because hydrogen has just one half the mass of deuterium it was preferentially lost from the planet. Two OH groups would eventually combine to give water plus oxygen by,

$$OH + OH \rightarrow H_2O + O$$

and the oxygen would combine with surface materials on Venus — for example, with sulphur from volcanic activity that would give sulphur dioxide and eventually the sulphuric acid that is now present in Venus's atmosphere. The net effect of all these reactions would be a gradual loss of water and an increase in the D/H ratio.

The overall D/H ratios in cool star-forming molecular clouds are the ratio for the Universe as a whole but the ratios are much higher in various hydrogen-containing molecules that form either icy grains, or icy coatings on silicate and iron grains, in cold molecular clouds and in the protostars that form within such clouds. In 2003, it was reported by H. Roberts and his colleagues that in the cold dense cloud L134N the ratio of doubly-deuterated ammonia, NHD_2, to normal ammonia, NH_3, is 0.05, meaning that 3.4% of the hydrogen in all the ammonia in that cloud is deuterium. In 2002, B. Parise and colleagues reported that in the protostar 16293E the D/H ratio in the ammonia it contains is more than 0.02 and the amount of methanol, CH_3OH, which is either singly or doubly-deuterated, i.e., CH_2DOH or CHD_2OH, actually exceeds that of normal methanol. Many other similar cases of high D/H ratios in icy grains in star-forming clouds have been found, some much higher than those reported here. The reason for this concentration of deuterium is a phenomenon known as *grain-surface chemistry*. If a deuterium atom in the gas falls on an icy grain it will dwell on the surface for a while. It may then change places with a hydrogen atom in one of the molecules of the grain since the change from hydrogen to deuterium in the molecule lowers its energy; physical systems are more stable with lower energy so the deuterium transfer will be permanent. Over the course of time the icy materials become increasingly deuterated. The consequent fall in the D/H ratio in the gas will be significant because although hydrogen accounts for the great majority of the mass the deuterium in the icy grains is enormously increased.

A protoplanet, formed by the capture-theory process, is initially at a very low temperature and remains low for the much of its collapse. During this stage the solid grains, of iron, silicate and ices, sink towards the centre of the collapsing planet, eventually to form an iron-silicate core with a surrounding mantle of ices, many of which contain deuterium-enriched hydrogen. The planet as a whole will be mostly hydrogen, with a deuterium deficiency, plus helium. When it has collapsed to near its final state the centre of the planet will be at a high temperature and the ices will vaporize and become part of the atmosphere, although concentrated at the base of the atmosphere because they consist of heavier molecules. Over a long period of time diffusion of gases will occur and the deuterium content will become more uniform in the planet.

33.6 The Leaning Planets

In general, at any instant, the evolving orbits of two planets with different semi-major axes, eccentricities and inclinations will not

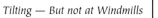
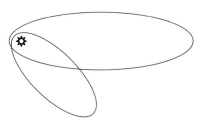

Figure 33.11 Two orbits intersecting in projection but not intersecting in space.

intersect in space. In projection onto the plane of the medium, as seen in Figure 33.11, they may appear to intersect but, because of their different inclinations, at the points of apparent intersection they will be separated in the direction perpendicular to the plane of the figure. However, because of different rates of precession for the planets, from time to time the orbits *will* intersect. Since the periods of the processions are of order 100,000 years, in general, for each pair of planets there are several orbital intersections during the period of orbital evolution.

The fact that orbits intersect, or nearly so, is a necessary condition for planets to closely approach each other but it is not a sufficient condition — they must also be in the vicinity of the intersection at the same time — a condition that will occur rarely but will happen occasionally. With notional starting orbits for the major planets, and their likely patterns of orbital evolution, calculations show that the probability of a close approach of any individual planet to some other during the period to be quite high — actually much more likely to happen than not to happen. Here we define 'close approach' as having a centre-to-centre distance within three times the radius of the larger planet, within which distance major tidal effects would occur. It should be noted that the planetary radii at the times of close interactions would have been somewhat greater than the final radii we now observe.

We now describe a mechanism by which the planetary spin axis tilt angles, given in Table 5.3, could have come about. Planets produced as condensations in a filament are subjected to tidal forces by the star and by other planets and the most likely direction of a spin axis is perpendicular to the planet's orbital plane. Figure 33.12 shows

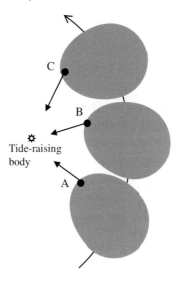

Figure 33.12 Tides on a planet (grey body) at successive times with tidal bulges at A, B and C.

the tidal effect on a planet due to another body, say the star. As the planet moves past the star its configuration changes, with the peak of the tide at three positions indicated as A, B and C. The peak does not point at the tide-raising body but lags behind because of inertia in the system; the material cannot react immediately to the changing forces that are acting on it. The arrows show the gravitational force acting on the tidal bulge and in each case it is pulling the bulge material into rotation around the centre of the planet in an anticlockwise sense. Eventually the planet returns to a near-spherical shape and the rotational motion in the tidal bulge is distributed throughout the body. The imparted rotation will be about an axis perpendicular to the orbital plane of the motion and the final rotation will be a combination of the original spin of the body with what was added. Since all the planets in a filament move closely in one plane, that of the star-protostar orbit, it might be thought that the spin axis due to tidal effects between the bodies would be almost perpendicular to that plane.

How then can the observed tilts of the planetary axes be explained — in particular the extreme tilt of Uranus? The answer lies

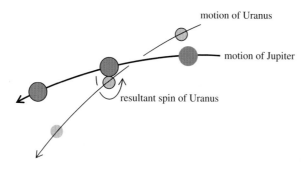

Figure 33.13 The relative motions of Jupiter and Uranus giving the spin axis of Uranus close to its orbital plane.

in the fact that orbits can intersect, or nearly so, and that they are not *precisely* coplanar. Figure 33.13 shows an early Uranus on an eccentric orbit with another more massive planet, taken as Jupiter, passing over it above its orbital plane. The tidal forces give a spin to Uranus about an axis almost in the plane of its orbit. Since Jupiter is much more massive than Uranus, its axis of spin would have been little affected. The smaller tilts of the other major planets can be explained in a similar way, although it is not possible to define a specific scenario that would explain them all precisely.

Uranus and Neptune are usually described as *ice giants* because they contain much less hydrogen and helium than the gas giants, Jupiter and Saturn, with proportionally a greater content of molecular species that would form ices at a low temperature. We now consider a scenario where an early Uranus is subjected to such high tidal forces that not only does it acquire an extreme axial tilt but it also loses a large part of its hydrogen-plus-helium atmosphere. Remember, in the early Uranus there was a layer of hydrogen-containing molecules, rich in deuterium, around the central core and this layer, in the deep interior of the planet, would have remained intact. Now, as over time diffusion restores the excess deuterium to the atmosphere, since the quantity of hydrogen with which the deuterium mixes has been reduced, the D/H ratio is increased in the final atmosphere. Calculations based on various assumptions give estimates of how much of the atmosphere has to be lost in the tidal interaction and a figure of about two-thirds both explains the final composition of Uranus and the final D/H ratio.

Now we can see why the question of the D/H ratio in major planets was introduced in Section 33.5. Both Uranus and Neptune are ice giants containing a high proportion of hydrogen-containing molecular species and both have higher than normal D/H ratios. The linkage of these two characteristics reinforces the plausibility of the proposed mechanism that simultaneously brings about both characteristics.

Part IX
The Biggish Bang Hypothesis

Chapter 34

The Terrestrial Planets Raise Problems

Notre Pére qui etes aux cieux Restez-y
Et nous nous resterons sur la terre.

Jaques Prévert (1900–1977)

34.1 The Problem

So far, we have produced a model that gives planets on orbits of the right size accompanied by satellites. Of course, we have not produced a detailed model of the Solar System as we see it now, with all planets having the right masses, in the right orbits and with their proper complement of satellites. That would be a tall order and it is no mean achievement just to show that the right general characteristics are feasible. Nevertheless, there is still a nagging doubt — does this type of model completely give what we should expect?

The protoplanetary condensations produced in the Capture Theory must correspond to eventual major planets; it is the total mass of the condensation, mostly gas, which gives a larger-than-critical Jeans mass that can collapse. The calculations of the type illustrated in Chapter 29 show that the protoplanets start on extended orbits with their initial motion away from the Sun. After 10,000 years or more, when they return to the vicinity of the Sun and reach perihelion, most of their material is in a compact central condensation that would be able to resist tidal disruption by the Sun. We have also seen that they would have had surrounding material in the form of an extended disk that would be a source of satellites by the accumulation of solid material. In view of that scenario, one wonders how it is that the small rocky terrestrial planets came about. An obvious idea that springs to mind is that since they ended up fairly close to the Sun, they would have heated up and lost their outer volatile material so that just a rocky core was left. However, observations of exoplanets show that large gaseous planets can exist at

317

distances of 0.04 au from stars that are very similar to the Sun, a distance which is about one-tenth of the orbital radius of Mercury. On the face of it, the capture-theory model seems likely to be able to produce only large gaseous planets so that some other mechanism for producing the rocky terrestrial planets appears to be necessary.

Before exploring this question of the origin of the terrestrial planets, we first restrict our attention to the Earth, consider its detailed composition and see what we can learn from that.

34.2 What Kind of Material does the Universe Contain?

Estimates have been made in various ways of the elemental composition of the Universe. They cannot be regarded as precise because they are derived indirectly by looking at the light from various sources — including the Sun and the Orion Nebula, for example — and judging the composition from the strength of the spectral lines corresponding to different kinds of atom. The elements that provide almost all the mass of the cosmos are listed in Table 34.1. The indication is that

Table 34.1   The main elelmental components of the cosmos

Element	Relative mass (Si = 30)
Hydrogen, H	31,800
Helium, He	8,840
Oxygen, O	354
Carbon, C	141
Nitrogen, N	51
Neon, Ne	69
Magnesium, Mg	25
Silicon, Si	30
Aluminium, Al	23
Iron, Fe	46
Suphur, S	16
Calcium, Ca	3
Sodium, Na	1
Potassium, K	2

something like 98.3% of the mass of the material is hydrogen and helium, with heavier elements accounting for the remaining 1.7%. About 4% of the mass of the heavier elements is silicon, the basis of the silicates that provide most, but not all, the stony material in the Solar System. Analysis of terrestrial, lunar and meteorite material suggests that the element silicon accounts for approximately one-fifth of the mass of stony material. That being so, then it seems that about 20% of the mass of the heavier-elements material in the cosmos goes into stones of one sort or another. The major part of the remaining heavier material is oxygen, nitrogen, carbon and inert gases, which exist in the form of gas or ices such as water ice (H_2O), solid carbon dioxide (CO_2), solid ammonia (NH_3) and solid methane (CH_4).

34.3 What Kinds of Material does the Earth Contain?

An important component of the Earth, and presumably of the other terrestrial planets, is iron, which exists both in the free form (combined with some nickel) in the core and as a component of minerals, such as olivine that forms the major part of the Earth's mantle. This division into free iron and mineral iron also occurs in stony meteorites of which the main type is the so-called *chondrites* because they contain *chondrules*, small glassy spheres. Chondrites are further divided into three sub-types.

- *Enstatite chondrites*, in which the iron occurs either as the metal or as troilite (iron sulphide) and which have a low magnesium: silicon ratio.
- *Carbonaceous chondrites*, where much of the iron occurs as magnetite (Fe_3O_4) and which contain volatile lighter elements and sometimes minerals, such as serpentine, with water of crystallization.
- *Ordinary chondrites*, by far the most common type, which are intermediate in properties between the other two.

In terms of iron content the ordinary chondrites fall into three classes — H (high iron), L (low iron) and LL (low iron — low metal). The iron content of these is shown in Table 34.2.

Table 34.2 Percentage of total iron and free iron in the three classes of ordinary chondrites

	H	L	LL
Total iron (%)	27	23	20
Metallic iron (%)	12–20	5–10	2

The usual interpretation of these three classes is that they represent different extents of loss of metallic iron during some high temperature episode in the Solar System and, if that is so, then the H class with the highest proportion of metallic iron must be the closest to the original material.

The proposition that we make here is that the Earth is the residue of some large body of material that originally contained the cosmic mix of elements and that the iron within it was divided between the free and mineral forms in the ratio 20:7 as suggested by the H class of ordinary chondrites with the highest free-iron content. There are some quite good models of the interior structure of the Earth from which it may be inferred that 35% of the Earth's mass is in the form of free iron in the core and if all the free iron in the original body had been retained then the total mass of iron, including that in minerals, would have been,

$$M_{Fe} = 0.35 \times \frac{27}{20} M_{\oplus} = 0.473 M_{\oplus}, \qquad (34.1)$$

where $M_{\oplus}$ is the mass of the Earth.

From this result we may infer that there is a deficiency of silicon in the Earth. On the basis that 20% of the mass of silicates is silicon, then the $0.65 M_{\oplus}$ of stony material in the Earth contains $0.13 M_{\oplus}$ of silicon. However, based on the ratio of iron and silicon in Table 34.1, the expected amount of silicon from (34.1) is

$$M_{Si} = 0.473 \times \frac{30}{46} M_{\oplus} = 0.308 M_{\oplus}. \qquad (34.2)$$

This indicates a loss of more than one-half (a fraction 0.178/0.308) of the original silicon in the form of silicates.

From Table 34.1, we may also estimate the mass of the original body produced by the Capture Theory from which the Earth is assumed to have been derived. Scaling the amount of iron indicated by (34.1) this mass would have been,

$$M_{\text{total}} = 0.473 \times \frac{41000}{46} M_\oplus = 428 M_\oplus, \qquad (34.3)$$

which is about $1.35 M_{\text{Jupiter}}$.

The numerical details of these calculations cannot be strongly defended since they depend on a number of assumptions. The estimates of the cosmic abundance of elements varies somewhat with the source, although not widely. The model of the Earth may be in error and some of the core may not be free iron but iron in combination with sulphur and oxygen. Again, it has been *assumed* that the H class of ordinary chondrites can be taken as the model of the composition of the original solid mass of material of which the Earth is the residue. Nevertheless, even given these uncertainties, it seems very likely that the material of the Earth was once associated with much more material, both in solid and gaseous form and that, in some way or other, there was a large loss of silicate relative to iron.

We now have a final question to answer. If the Earth was originally part of the solid component of a body of greater than Jupiter mass, then how was it derived from that body?

Chapter 35

A Biggish Bang Theory:
The Earth and Venus

Thou stand'st unshook amidst a bursting world

Alexander Pope (1688–1744)

35.1 A Very Close Encounter of a Planetary Kind

In Chapter 33, we found that the initial protoplanet orbits were highly eccentric, inclined to the ecliptic by, at most, a few degrees and that they underwent precession. It was also shown how near-passages of pairs of major planets could give rise to tilts of their spin axes and, sometimes, even the loss of considerable parts of the gaseous component of one of them, so converting it into an ice giant. Now we look at an even closer interaction — in fact, a collision.

In Section 33.7, it was explained that if the eccentric, inclined and extended orbits of two planets underwent precession at different rates then, from purely geometric considerations, from time to time their orbits will intersect in space. The intersection of orbits does not guarantee that a collision will take place; the planets must also reach the point of intersection at the same time. Given the characteristics of two planetary orbits and the radii of the planets it is possible to calculate how long it would take on average for the two planets to collide. Examples of such calculations were given by John Dormand and me in 1977. The calculated average times for pairs of planets to collide turn out to be greater than the round-off times for the planets; of course, once the planets have rounded off into well-separated near-circular orbits then no collisions can take place. However, it is only the *calculated average* time that is greater than the round-off times and in any real physical situation the *actual* times for a particular collision may sometimes be less than the calculated average time.

Taking into account all possible pairs of planets that might collide, it was estimated that the probability of a planetary collision in the early Solar System is reasonably large — of order 0.1 or so, just to quantify it. The idea that planetary collisions can occur received support from a NASA Spitzer Space Telescope observation in 2009 of evidence of a planetary collision in the vicinity of the young star HD172555 (age 12 million years) within the last few thousand years.

As an event, a planetary collision lies well outside the range of everyday experience in terms of the energy that would be generated. Yes, we see collisions on Earth, sometimes due to tragic accidents, and we have seen films of the most energetic man-made phenomena, thermonuclear explosions. In February 2013, a tiny asteroid, of mass 10 tonnes, landed in the Chelyabinsk region in the Ural Mountains in Russia, causing injuries to 1,000 people, mainly due to shattered glass. It is thought that a much larger asteroid, colliding with the Earth some 65 million years ago, brought about the extinction of the dinosaurs; the energy released in such an event would have dwarfed even a man-made thermonuclear explosion. However, that asteroid was probably no more than a few kilometres in diameter, a trivial object compared to a major planet. We can carry out calculations to indicate what would have been the energy production from the collision of two major planets, but it is difficult to *imagine* such an event. The energy released in such a collision would be at least 10^{18} (a million million million) times as much as in the most powerful man-made thermonuclear explosion. Another comparison is that it would release in one event, of duration measured in hours, the total energy output of the Sun over a three-year period.

35.2 The Colliding Planets

Following the example of an early attempt to study the effects of a planetary collision by John Dormand and me in 1977, the early Solar System is taken as having originally had six major planets, the present four plus two others that we call Bellona and Enyo.[1] Something

---

[1] Bellona was the Roman godess of war and Enyo her Greek counterpart. There is also an asteroid named Bellona.

we know from the measurement of exoplanet masses, as illustrated in Table 17.1, is that planets of well above Jupiter mass are quite common. If we take at least one of the colliding planets as considerably more massive than Jupiter then many disparate features of the Solar System can be explained as by-products of the collision. There is no way of knowing what the actual planetary masses were, so all that can be done is to show that with some postulated masses an outcome is obtained that explains features of the Solar System — although other pairs of masses might also give acceptable outcomes.

The two model planets used to study the collision had masses of $799 M_\oplus$ for Bellona and $598 M_\oplus$, for Enyo, where $M_\oplus$ is the mass of the Earth — about 2.5 and 1.9 times the mass of Jupiter respectively. In creating models of these early planets, the fact that in their central regions the heavier material — iron and silicates — would not be completely separated, was taken into account. In the Earth such separation is complete but it has been estimated that this would have taken about ten million years from its initial formation. Each of the model planets had four separate layers — a central core made of a 40:60 mixture of iron and silicate, a mantle with a 85:15 mixture of silicate and iron, a thick layer of 'ices' and an enveloping hydrogen-helium atmosphere (Figure 35.1). Ideally in the central region, covering the core and mantle, the iron:silicate ratio should smoothly decrease with distance from the centre but in making models one must compromise between strict reality and computational practicality.

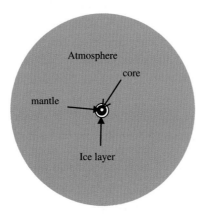

Figure 35.1 A model planet with four layers.

Table 35.1   The characteristics of the colliding planets

Planet	Bellona	Enyo
Mass ($M_\oplus$)	799	598
Radius (km)	91,520	86,470
Central density (kg m^{-3})	176,500	146,500
Central temperature (K)	85,000	74,000
Mass of core ($M_\oplus$)	3.00	2.50
Mass of mantle ($M_\oplus$)	12.00	10.00
Mass of ice ($M_\oplus$)	6.00	5.00

We should also note that when we refer to 'ice' we mean materials such as water (H_2O), methane (CH_4), ammonia (NH_3) and carbon dioxide (CO_2) that will have been solid icy grains in the original material that produced the planet and in which any hydrogen will be deuterium rich (Section 33.6). These materials may actually be in liquid or gaseous form but we shall still call them 'ice'. In a real planet there would not be sharp divisions between the layers. In particular there will be ice-impregnated silicate regions and silicate-impregnated ice regions. The ice-atmosphere boundary would also certainly have been blurred with mixing of atmospheric material with ice-material vapour. The ice layer would have contained a great deal of hydrogen and this was taken with a D/H ratio of 0.01, well within the limits of observation. However, the D/H ratio of the atmosphere was taken just at the cosmic level of 2×10^{-5}. Models of these planets, with a particular composition of iron, silicate, icy materials and gases, were made using the standard equations that describe the way that pressure, density and temperature vary with distance from the centre of a large astronomical body. As would be expected, the conditions in these two giant planets are more extreme than in Jupiter. The characteristics of the model planets, including the masses contained in each of the layers, are given in Table 35.1.

35.3 The Collision

Once the physical properties of the model planets had been determined they were then transformed into a form that made them suitable for a smoothed-particle-hydrodynamics simulation (Section 29.3). A spherical distribution of points, covering the volume of each planet was set up with associated masses and temperatures that reproduced the deduced distribution of density and temperature and with physical properties appropriate to the type of material they represented. As will be seen later, the main interest in this simulation is to follow the behaviour of the central layers so, to increase the resolution of the model where it was required, the density of SPH points was greatest at the centre of each planet and reduced with distance from the centre.

In the simulation the planets were started some distance apart and then given velocities that caused them to approach each other and eventually collide. The progress of the collision is illustrated in Figure 35.2, starting from just before the planets came into contact. As the collision developed so the region of highest temperature, which steadily increased in value, moved towards the centres of the planets. The first ice region it reached is that of Enyo and when it did so the temperature had reached several million K.

When the high-temperature region reached the ice layer in Enyo, nuclear reactions took place involving two deuterium nuclei — as happens in brown dwarfs (Section 29.4). Reactions involving heavier nuclei require much higher temperatures, of order 10^8 K, in order for them to occur. The deuterium reactions took place explosively and, due to both the high concentration of hydrogen-containing molecules plus a D/H ratio of 0.01, the temperature generated quickly rose to about 10^8 K, at which temperature further reactions took place involving heavier elements present in both the ice and the silicate material that accompanied the ice. The final temperature generated by all these nuclear reactions was several times 10^8 K and the energy produced was sufficient to disrupt the planets and expel most of their material from the Solar System.

The stages in the collision are:

(a) Time = 0.   The planets are just about to make contact.

(b) Time = 590 s.   The planets are merging but no nuclear explosion has taken place.

(c) Time = 1,326 s   A nuclear explosion has occurred and the planets are expanding outwards.

(d)–(i) Time = 2,505–8,609s   Most material of both planets is violently dispersed but the residues of two cores have survived and are moving apart.

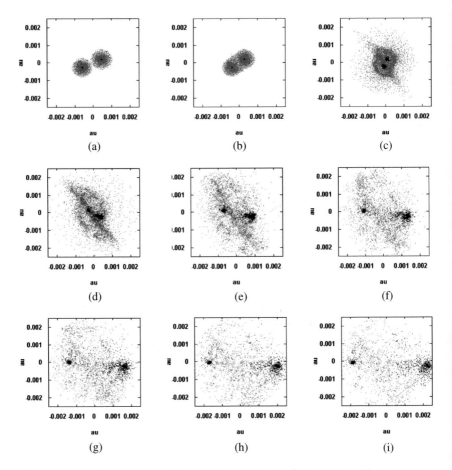

Figure 35.2   The progress of the collision. (a) $t = 0$, (b) $t = 590$ s, (c) $t = 1,326$ s, (d) $t = 2,505$ s, (e) $t = 3,917$ s, (f) $t = 5,336$ s, (g) $t = 6,415$ s, (h) $t = 7,597$ s, (i) $t = 8,609$ s. Points corresponding to the core are black; other points are grey.

The proposal that we now consider is that the surviving residues of Bellona and Enyo are, respectively, the embryonic terrestrial planets Earth and Venus. It will not be possible to show that the masses and final orbits precisely match those of the larger terrestrial planets, but we will show that two terrestrial-type bodies will end up orbiting in the terrestrial region of the Solar System.

35.4 Orbital Considerations

The course of the collision was calculated as though the bodies were isolated but they would actually have collided in the presence of the Sun. Treating the collision scenario as isolated is quite acceptable because, within the comparatively small region of the simulation, the perturbing influence of the Sun would have been extremely small. To illustrate this, it should be noted that the maximum distance between the colliding planets is very little greater than the Earth–Moon distance and the Sun has very little influence on the orbit of the Moon round the Earth. However, if we wish to find the orbits of the planets around the Sun, both well before the collision, when they were far apart, and well after the collision, when again they were far apart, we have to fix the position of the collision, and the velocity of the centre of mass of the colliding planets, relative to the Sun. We cannot know what these quantities were, so, once again, all that can be done is to show that reasonable results are found for some configurations. Various conditions were tried, many of which gave acceptable results. For one example, collision at a distance of 0.400 au from the Sun with the initial semi-major axes and eccentricities (0.433 au, 0.149) for Bellona and (0.465 au, 0.733) for Enyo gave final residue orbits (2.759 au, 0.873) for Bellona and (0.873 au, 0.971) for Enyo.

The collision took place while the resisting medium was present and orbits were still being modified. In the numerical studies of orbital evolution in a resisting medium, described in Chapter 30, it was found that, for a given medium and initial orbit, the timescale of orbital evolution *decreased* as the mass of the planet *increased*. Both Bellona and Enyo were more massive than the present planets of the Solar System and hence their orbits would have evolved more quickly, taking them close to the Sun where a collision would be more probable due to the restricted region within which the planets moved. The medium was still substantial at this stage and we must now consider how the post-collision orbits of the residues may have evolved to their final states.

In Figure 30.5 we showed the effect of a resisting medium on an elliptical orbit. At perihelion both the semi-major axis and the eccentricity were reduced. At aphelion the eccentricity was reduced but the semi-major axis was increased. The rounding-off of the orbit is a common feature at both positions but the relative importance of the effect on the semi-major axis depends on the radial distribution of density in the medium. For a medium that decreases steadily with distance from the Sun the perihelion effect will be strongest but it is possible to have a medium like that shown in Figure 30.7 where for orbits with an aphelion in the region of greater density the effect there will dominate. For the post-collision orbits we have found here, the most likely outcome is that the final evolved circular orbit for the Bellona residue (Earth) will be more than 1 au from the Sun and that for the Enyo residue will be less than 0.72 au, the value for Venus. We do not have a match of orbits but what we do have is two planetary residues in circular orbits in the terrestrial region of the Solar System.

35.5 Mass and Composition Considerations

The final masses of the residues cannot be defined with certainty because it is not clear where to take the boundaries. However, the centre of mass (CoM) of each residue can be found reasonably well and the masses contained within different distances from each CoM at the stage represented by Figure 35.2(i) can be found. The result of doing so is shown in Figure 35.3.

The Bellona residue is much more compact and contains a mass of $2.5M_{\oplus}$ within 20,000 km, just over three times the radius of the Earth. The Enyo residue contains $1.5M_{\oplus}$ within 100,000 km, which is about one-and-a-half times the radius of Jupiter. Both these residues would have a free-fall time to reach high density of hours or days (Equation (27.1)), although their high temperatures will produce pressures that slow down their collapse. Both residues are more massive than the terrestrial planets they are meant to represent but this reflects the characteristics of the model planets that were used to simulate the collision. The core of Bellona had a mass of $3M_{\oplus}$ and

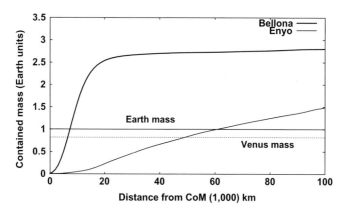

Figure 35.3 The variation of the contained mass with distance for the residual cores.

that of Enyo $2.5M_{\oplus}$ and the residues, which are of lesser mass, are dominantly core material, with most of the mantle and all the ice material from the original planets having been widely dispersed by the nuclear explosion. Once more we have not matched the detailed characteristics of the terrestrial planets, this time in terms of mass, but once again we have characteristics that have the right ballpark values, within a factor of two, more-or-less.

The residues, like the cores from which they are derived, have a composition close to 40% iron and 60% silicate (they also include a little mantle material), which will be reflected in the final terrestrial planets they form. This is a somewhat greater proportion of free iron that is found in the larger terrestrial planets (~ 35%) but, in the loss of most of the mantle material which was dominantly silicates, we do have here an explanation for the deficiency of silicates in the Earth.

While the final composition of the larger terrestrial planets has just been described in terms of discrete compositions in the core and mantle, the actual position would have been a fall in the ratio of iron:silicate with increasing distance from the centre. While the composition of the combined core-plus-mantle would have given close to the cosmic iron:silicate ratio the residues forming the larger terrestrial planets would have come mainly from the inner parts of the original solid regions of Bellona and Enyo and hence be iron-rich and silicate-poor. The fact that Earth has a slightly higher proportion of iron as compared to Venus would then be a reflection of the greater mass of Bellona that would have given somewhat greater segregation.

35.6 Summary and Comments

It has been shown that the collision of two planets, with orbits corresponding to evolved capture-theory orbits and with masses in the range observed for exoplanets, can give rise to stony-iron residues of mass similar to those of terrestrial planets orbiting in the terrestrial region of the Solar System. This suggests that Venus and the Earth may have formed in this way. The residues will be enriched in iron, and hence depleted in silicates, which explains the difference between the composition of the Earth and that of the general material of the cosmos. Although the simulation described here has been carried out for a particular pair of planetary masses, similar calculations with pairs of masses, in Earth units, (452, 169), (618, 116) and (990, 333) have yielded similar outcomes.

If the idea of a planetary collision was just an *ad hoc* explanation for the larger terrestrial planets, and nothing else, then it would still be plausible since it is based on sound science and observations. However, as we shall see, it explains much more.

Chapter 36

Behold the Wandering Moon

To behold the wandering Moon,
Riding near her highest noon,
Like one that has been led astray

John Milton (1608–1674)

36.1 Orphans of the Storm

All the major planets have considerable satellite families and Bellona (that gave the Earth) and Enyo (that gave Venus) were not exceptional in this respect. Indeed, since Bellona has been taken with 2.5 times the mass of Jupiter then it should have had satellites considerably more massive than Ganymede, the most massive of the Galilean satellites. Given that satellites once existed for the two colliding planets then the question arises of what became of them following the planetary collision.

There are only five possible outcomes for each of these satellites:

(i) It could have been retained by the residue of the original parent planet.
(ii) It could have changed parent and been captured by the other residue.
(iii) It could have gone into an independent orbit around the Sun.
(iv) It could have moved fast enough to have left the Solar System altogether.
(v) It could have been broken up by the collision event and its debris dispersed.

The first four outcomes can be simulated by computer calculations. Possible interpretations of the Moon's association with the

Earth are that it was either an original satellite of Bellona that was retained by the Bellona residue (the Earth) or that it was a satellite of Enyo that attached itself to the Bellona residue. The former option is the more likely and the one we shall assume here for the purpose of discussion.

Other satellites, with other destinies, are considered later.

36.2 The Pre-Collision Moon

In Chapter 32 we described the process by which the satellites formed. In the final stage of the process satellitesimals accumulated to form a satellite, which gradually grew from a small initial assemblage of satellitesimals to the final body. When a satellitesimal fell on to the growing body its kinetic energy would have been converted into other forms of energy, much of it as heating of the region on which it fell. As the mass of the embryonic satellite increased so the kinetic energy of the arriving satellitesimals, influenced by the gravitational attraction of the growing body, would have increased and so would the generated local temperature. The expected form of the temperature profile in the final body was found in 1973 by Toksöz and Solomon who also took other factors into account. One of these was that when a body landed on the growing satellite some of its energy would be converted into shock waves that would travel into the interior of the satellite and deposit heat energy there. Another factor was that the outer surface of the growing satellite would always have been radiating heat outwards and so cooling. There would also be differences in the initial thermal profile, when the satellite had completely formed, which would depend on the rate at which it grew; the longer it took to grow the greater would have been the loss of heat by radiation during the growing process. A typical initial profile for the Moon is shown in Figure 35.1; particular attention should be paid to the non-linear depth scale (horizontal axis), designed to give more resolution close to the surface so that the variation of the thickness of the solid layer with time can better be seen. Another feature of interest is the *solidus line*, above which material is fluid and below which it is solid.

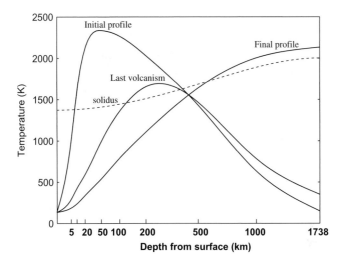

Figure 36.1 The evolution of temperature profiles in the Moon.

As time passed so heat was transported internally and also radiated from the surface, the net result being that the location of maximum temperature migrated towards the centre and the solid layer at the surface became thicker. The calculated profiles are shown for the time at which the surface layer becomes so thick that volcanism ceases and also for the present time. At the surface of the Moon there would have been a crust of lighter silicates that rose to the top of the molten mix of silicates and iron and, as time passed so a core of denser iron would also have formed; from the present physical characteristics of the Moon the core is estimated to have a radius of about 380 km.

A further factor that must be taken into account is that the Moon, originally in a fluid or plastic state and hence easily distorted, formed within the gravitational field of Bellona. Once it had settled down, with its spin period equal to its orbital period so that it always had one hemisphere facing Bellona, it would take up a profile as seen in Figure 36.2. An obvious feature is that the Moon has a slight pear shape (exaggerated somewhat in the figure) with the pointed end towards Bellona; the present Moon has this feature but with the pointed end towards the Earth. The other feature is that the lighter-density crust, shown yellow, is thicker on the side facing Bellona than on the reverse side.

To Bellona

Figure 36.2 The profile of the pre-collision Moon, showing the differential thickness of crust on the near and far sides. The distortion of the Moon and thickness of crust are exaggerated for clarity of presentation.

We now consider how the Moon in this state, with synchronous spin and orbital periods, would be affected by the collision.

36.3 A Lopsided Moon

Due to its synchronous spin and orbit the Moon always presents one face to the Earth. The Moon is not alone in this respect — it is a general characteristic of satellites that they orbit with one face permanently towards their parent bodies. For this reason, throughout history until recent times, we have only been able to observe that side of the Moon that faces the Earth and it was generally assumed that what we could see was a good sample of the Moon's surface as a whole. Actually, because the Moon's orbit around the Earth is in the form of an ellipse, with eccentricity 0.056 (Figure 2.2), it appears to oscillate slightly over the course of a lunar month, a motion known as *libration*. This means that we can actually see 59% of the Moon's surface, although the view at the edges is very distorted.

In 1959 a Lunik spacecraft, launched by the then Soviet Union, made an historic journey round the back of the Moon and sent back photographs to Earth. The pictures were of poor quality but they were good enough to be sensational. The rear side of the Moon was completely different from the side we could see from Earth. The Moon was lopsided not only because of its shape but also in terms of its surface features. Later, and higher-quality, images starkly illustrated the difference of the two hemispheres.

The photograph in Figure 36.3, taken by a later Soviet Union spacecraft, shows part of the near side of the Moon on the right and

Figure 36.3 The near side (on the right) and far side of the Moon in the same image.

part of the rear of the Moon on the left. The difference between the two sides is very clear. However, it is not that the *kinds* of terrain they contain are different but rather that the *proportions* of the various kinds of terrain are different. The various kinds of terrain are well illustrated in the full-Moon picture shown in Figure 36.4, which also shows the numbered Apollo spacecraft landing sites. The two major types of surface feature are the darker regions, which are *mare basins*, and the lighter *highland regions*. Mare basins were formed by large projectiles that struck the Moon and hollowed out what were, essentially, enormous craters. The early Moon had hot fluid material not far below its solid crust. The basins corresponded to weakened and thinned regions of the crust and lava was able to well up and to fill the basins. Subsequently, these areas were volcanically active for several hundred million years, as indicated by overlapping solidified flows of lava that can be seen in some photographs. There are relatively few smaller craters within the mare basins. This is because most craters there were produced very early in the history of the Solar System when there were many projectiles and they became obliterated by the lava flows that later covered them. By contrast the highland regions, representing areas that did not have mare basins formed within them and hence had no episodes of volcanism, and

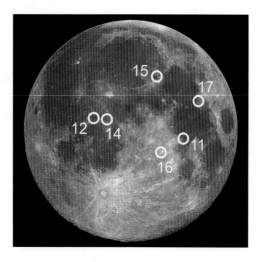

Figure 36.4 Near side of the Moon showing the Apollo landing sites.

the record of the intense bombardment, that produced a profusion of craters endlessly overlapping each other, is preserved.

The difference between the two hemispheres is simply stated. The side that faces the Earth contains many large mare basins to the extent that the terrain is dominated by these features. By contrast the far side is almost exclusively highland terrain with very few small mare basins. The question arises — "why is this?"

36.4 The Lopsided Moon — An Answer and a Question

An obvious explanation for the hemispherical asymmetry of the Moon would be that, for some reason or other, large projectiles fell only on the side facing the Earth. One early idea put forward was that the Earth's gravity acted in such a way that it focussed projectiles onto the Moon's near-surface. This theory was not supported by analysis but, more importantly, measurements from spacecraft showed that it was wrong. Spacecraft travelling round the Moon had radar equipment on board and by timing radar signals reflected back from the surface they were able to map the surface undulations. What this showed is that there *are* large basins on the far side of the Moon but they simply did not have lava inflows to fill them.

So, bang goes that explanation; the Moon was bombarded equally all over its surface.

The various Apollo missions that landed on the Moon left behind numbers of measuring devices including seismometers. These are the instruments used on Earth for recording earthquakes and, since earthquake waves travel through the Earth, the information from them enables the internal structure of the Earth to be determined. The Moon is a very quiet body and such moonquakes as there are have a tiny fraction of the energy of a normal earthquake. Nevertheless, they can be recorded since the seismometers left on the Moon are very sensitive. However, the seismometers can also pick up other signals — for example, due to meteorites striking the Moon's surface. In one Apollo mission, part of a Saturn rocket was deliberately targeted on to the Moon to give a good seismic signal. The outcome of these measurements is that we now know something about the Moon's internal structure. For our present purposes all we need to know is that the average thickness of the Moon's crust is about 60 kilometres and, more significantly, that the crust on the near side is about 25 kilometres *thinner* than that on the far side — as illustrated in Figure 36.5.

Here then is the answer to the problem of the asymmetry of the Moon. When the Moon first formed all its outer regions would have been molten. The less dense crust material at the surface would have quickly cooled and solidified and was thinner on the near side.

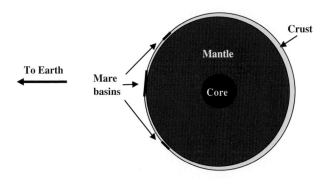

Figure 36.5 The difference of thickness of the crust on the two sides of the Moon (not to scale).

Later, over a lengthy period of time, very large projectiles occasionally landed on its surface producing basins. On the near side the solid shell was thin enough for lava to penetrate from below and form the mare. On the far side the lava was too far below the surface to fill the basins.

A very neat answer — but now comes the question, "Why is the crust thinner on the near side?" This question is all the more pertinent since, as we have already noted, the molten Moon, with a layer of less dense crustal material at its surface while settling down in the presence of a planet, which we take as Bellona, should have had a *thicker* crust on the near side, as shown in Figure 36.2.

36.5 The Collision to the Rescue

Let us now look at the planetary collision from the Moon's point of view, taking Bellona as its original parent. The slightly pear-shaped Moon orbited Bellona with one face pointing towards its parent with a period that cannot be known but was probably between a few days and a month. Now, from a direction that did not bring it near the Moon, Enyo crashed into Bellona. Within minutes mayhem broke loose. Debris moving at 100 km s^{-1}, or more, collided with the Moon. It is important to understand what happens when projectiles land on a solid body, such as a satellite, something which is illustrated in Figure 36.6. When the projectile lands it shares its energy with more than its own mass of surface material, some of the energy appearing as heat and the remainder as kinetic energy of surface material that flies up off the surface. If the projectile falls on the surface with a speed not much greater than the escape speed from the body (theoretically it cannot be less), like a meteorite falling to Earth, then none of the debris has greater than escape speed so it returns to the surface. The net result is that the *projectile is accreted by the solid body*. By contrast if the projectile lands with a speed that is very much greater than the escape speed then the debris, or a large part of it, can have greater than the escape speed and the *solid body is abraded by the projectile*. In the present case the escape speed from the Moon is about 2.4 km s^{-1} so the projectile landed with 40 or

Slow projectile   Fast projectile

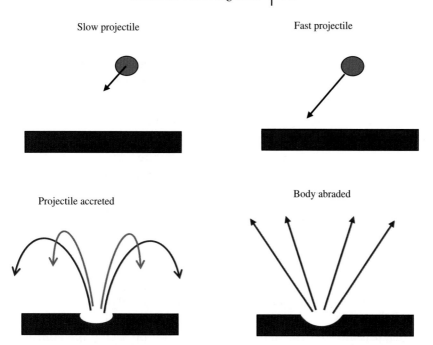

Projectile accreted   Body abraded

Figure 36.6   The effect of slow and fast projectiles falling on to a satellite.

more times the escape speed and, hence, the bombarded surface was abraded.

The Moon was spinning on its axis quite slowly, with the period of its spin equal to the period of its orbit, and in the period of hours during which the collision event was taking place it would have spun very little. The rear side of the Moon was shielded from the debris and would have been essentially unaffected by the collision. The near-side material that was removed caused the thinning of its surface and calculations show that, given the total mass of solid debris that would have been released by the collision, the removal of a few tens of kilometres thickness of crust of the near-side is a plausible outcome.

We see that the collision really provides the answer to two questions. The first is that of how the Earth *acquired* a satellite so large in relation to its own mass. The answer given here suggests that the question is actually posed in the wrong way. In a sense the Moon was *always* associated with the Earth, but the Earth itself is just the residue of a onetime major planet, Bellona, most of which was dispersed. The second question relates to the asymmetric appearance of the Moon that depends on the crust being thinner on the near side. A collision provides a logical reason for this. The abrasion of the surface was sufficient to convert a hemisphere of crust that was thicker than average into one that was thinner than average.

36.6 A Brief History of the Moon

To summarize the events that have been described above, the stages in the history of the Moon are as follows:

1 Bellona is formed in a capture event and the Moon, together with other satellites, is derived from the surrounding disk of material.
2 The collision takes place, with the Moon being retained by the Bellona residue (the Earth).
3 Debris abrades the hemisphere of the Moon facing the collision, converting what was originally the thickest part of the crust to become the thinnest part of the crust.
4 The Solar System is now full of orbiting debris that occasionally collides with other bodies (mainly major planets) and is absorbed. From time to time, over the course of the next hundred million years or so, projectiles fall on the Moon, the largest ones producing basins distributed all over its surface.
5 Basins on the near side, where the crust is thinner, fill with lava from below and are also regions of volcanism over several hundred million years. Craters produced in these regions by smaller projectiles are obliterated by these flows. Craters in highland areas remain as permanent scars on the surface.
6 As the Moon cools so the molten region moves deeper into the interior and eventually volcanism ceases. The Moon takes on its present form, unmodified except for collisions that continue to produce craters, although at a much reduced rate.

Chapter 37

Fleet Mercury and Warlike Mars

Rise from the ground like a feather'd Mercury

William Shakespeare (1564–1616), *Henry IV, Part I*

37.1 Mars as an Orphan

If the Earth and Venus cannot readily be explained as having been formed directly within the filament produced by the capture event then this is even more so for the two smaller terrestrial planets, Mars and Mercury. Mars has about one-ninth, and Mercury about one-twentieth, the mass of the Earth. To understand how these small planets fit into the general pattern of small bodies within the Solar System it is instructive to look at the terrestrial planets and the seven largest satellites strung out according to their uncompressed densities, the densities they would have if they were not compressed by their own gravitational forces (Figure 37.1).

When the smaller bodies are looked at in this way, the grouping suggests that Mars could readily be associated with the rocky satellites rather than with the larger terrestrial planets. We will leave consideration of Mercury aside for the time being. Well then, how possible is it that Mars was a satellite of one of the colliding planets that was released into a heliocentric orbit that, after partial rounding-off and decay, ended up where it is today? It is certainly much more massive than the present satellites. The masses of the largest satellites and of the terrestrial planets are listed in Table 37.1, with their masses given in Moon units, and it is clear that the two smaller planets are considerably more massive than Ganymede, the most massive satellite. Nevertheless, they are much closer in mass to the larger satellites than to the larger terrestrial planets. Bearing in mind that the postulated Bellona had 2.5 times the mass of Jupiter it could,

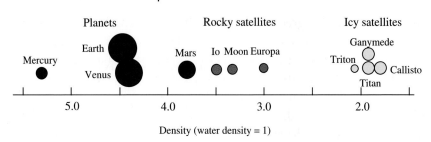

Figure 37.1 Uncompressed densities of the terrestrial planets and the largest satellites.

Table 37.1 The masses of the terrestrial planets and the largest satellites

Body	Mass (Moon units)
Earth	81.3
Venus	66.3
Mars	8.73
Mercury	4.49
Ganymede	2.04
Titan	1.93
Callisto	1.45
Io	1.21
Moon	1.00
Europa	0.66
Triton	0.30

quite reasonably, have had satellites more massive than those of any of the other planets.

Apart from the theoretical possibility that Mars could have made the transition from being a satellite to being an independent body orbiting the Sun, is there any other clue that a collision could have been involved? Yes, there is. In a similar way to the Moon, Mars shows hemispherical asymmetry as is illustrated in Figure 5.7. This image of Mars shows its topography with height represented by the colours of the spectrum — red representing the highest regions and blue representing the lowest. The red-to-yellow highland regions in the south and the green-to-blue northern plains show clearly the division between the two hemispheres. The division is marked by a scarp (a steep slope), 2 to 3 kilometres high, that runs at an angle of about 35° to the equator. The northern plains are covered with volcanic lava and contain comparatively few craters while the southern highlands, like the highlands of the Moon, are heavily cratered. The deep-blue region in the south is the Hellas Basin, 1,800 kilometres in diameter and 3 kilometres below the average height of the Martian surface.

These features can be explained in terms of Mars having been a satellite that was heavily bombarded on one hemisphere by the debris from the colliding planets. In the case of the Moon the crust was thinned on the near side to the extent that subsequent bombardment by huge projectiles enabled mare basins to fill with magma from below. However, for Mars the removal of material was sufficient to give penetration close to, or even into, the sub-surface molten material and extensive lava flooding of the whole region took place. This could have been due to the removal of a great thickness of crust or to the presence of molten material closer to the surface (likely for a larger body) or even both. The southern highland region then represents the unaffected hemisphere, but this crust was penetrated by a massive projectile some time later to produce the Hellas Basin. There is another large impact basin in the south, Argyre, which is about 1,000 kilometres in diameter.

There is evidence that, early in its history, Mars had a more extensive atmosphere than now and also had flowing water on its surface. It probably had a thin covering of ice, like Jupiter's Europa, and much of this ice would have been melted and vaporized by the effects of the collision and its aftermath to contribute to a dense atmosphere. There would have been two main results of such an atmosphere. Firstly, due to the blanketing action of an atmosphere rich in carbon dioxide and water vapour (both very effective greenhouse gasses) the planet would have been much warmer than now so that liquid water could have existed on its surface. In addition, with so much water vapour in the atmosphere, precipitation in the form of rain would have occurred and, consequently, running water to

Figure 37.2 Dried up water channels on the Martian surface.

form the channels that resemble riverbeds that are seen on the surface (Figure 37.2). Some of that water is now locked up in the Martian poles and another part is present as permafrost below the Martian surface, something that has been verified by spacecraft observations. Most significantly, in August 2011 there came evidence of the transient occurrence of liquid water on the Martian surface even now (Section 5.4: Figure 5.5).

A proposed history of the formation, and eventual loss, of a Martian water-based climate was given by Tony Connell and me in 1983.

An important difference between the hemispherical asymmetries of Mars and the Moon is their relationship to the spin axes or, alternatively, to the equator. The spin axis of the Moon did not change relative to surface features subsequent to the collision because it became locked to the Earth in the same configuration as it had previously been locked to the complete planet, Bellona. Consequently the plane dividing the two hemispheres contains the lunar spin axis. Mars, which went into an independent orbit around the Sun, was larger, much hotter and more fluid than the Moon. The remains of the solid crust floated on a liquid mantle and motion of the crust relative to the bulk of the planet could readily take place. The American astronomers, J.A. Lamy and P. Burns, produced a theorem showing that for a body in which energy was dissipated, as would happen due to fluid flows within Mars, and for which angular momentum remained constant, as would be so for an isolated body like Mars, the material of the body would arrange itself to be as far from the spin axis as possible. In 1983, Tony Connell and I showed that the present arrangement of surface features on Mars — both high features such as Olympus Mons and low features such as the Hellas Basin — closely satisfies the Lamy and Burns condition.

37.2 Mercury as an Orphan

Although Mars can readily be associated with rocky satellites by considering its density, that is certainly not true for Mercury. Indeed, since the density of the Earth is influenced by self-compression, an effect that is negligible for Mercury, the basic uncompressed density of Mercury is considerably greater than that of the Earth, as is seen in Figure 37.1. The reason for Mercury's high density is due to its relatively large iron core that accounts for about 75% of the radius of the planet. The internal structure of Mercury, compared with that of Mars, is shown in Figure 37.3.

The iron core of Mercury is about the same size, perhaps a little larger, than that of Mars. In the past it has often been suggested that Mercury was originally a larger body that was impacted by a large projectile that stripped away a major part of its mantle to give what

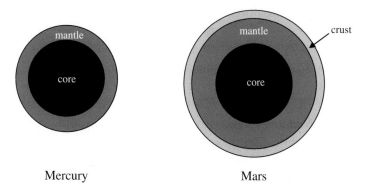

Figure 37.3 The internal structures of Mercury and Mars.

it is now. The planetary collision hypothesis offers a scenario in which this could have occurred, although not necessarily as a result of a single large projectile. If a satellite similar in size to Mars was very close to the collision, i.e. in an orbit close to one of the colliding planets, or, alternatively, was directly in the path of the highest intensity of debris, as indicated in Figure 35.2 (c), (d) and (e), and had somewhat more than 50% of its mantle stripped away, then what was left would have a similar composition to Mercury — something I proposed in 2000. In the case of the Moon and Mars the integrity of the satellite was not totally destroyed by the abrasion process. There had to be some rearrangement of material, mainly internal, to restore equilibrium but, for the most part, the final body after abrasion would have clearly resembled the original one. In the case of Mercury that would not have been so. Immediately after the removal of a large part of the mantle the residue would have been an amorphous mass of material, mostly iron on the side facing the collision and mostly silicate on the other side. Gravitational forces would then have reassembled it into a near-spherical form with an iron core and a reduced mantle, as it is today (Figure 37.4).

The surface features of Mercury, which have been likened to those on the Moon, can be interpreted in relation to this kind of catastrophic origin. The number density of craters is less than on the Moon and there are smooth areas between the craters — which shows that the early surface was fluid for a proportion of the time when heavy bombardment of solar-system objects was taking place. A very large impact feature, the Caloris Basin, is shown in Figure 5.8. The impacting body created a series of rings that are clearly seen — produced like the ripples in a pond when a stone is thrown into it.

Mercury has a very eccentric orbit, with $e = 0.206$, and its spin period, 58.65 days, is precisely $\frac{2}{3}$ of the orbital period, 87.97 days. This means that there are two regions, at diametrically opposite locations of Mercury, which are sub-solar, and become very hot, at alternate perihelion passages. The reason for the name Caloris Basin (*calor* is the Latin for *heat*) is that it is one of these two regions, and at the diametrically opposite location there is a region of very disturbed surface called the *chaotic terrain*. The usual interpretation of

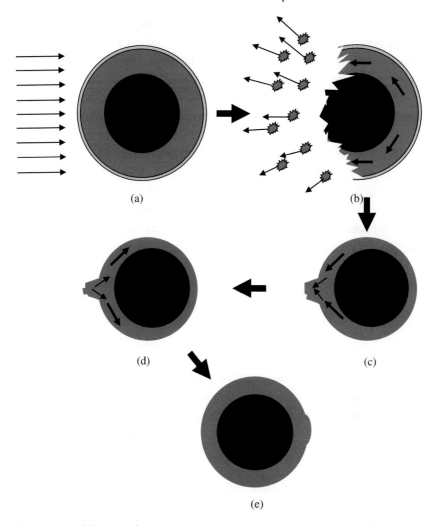

Figure 37.4 (a) A rain of projectiles approaches the exposed hemisphere of Mercury. (b) Most of the mantle of the exposed hemisphere is stripped away. (c) The mantle flows round from the unaffected hemisphere and, due to overshoot, material piles up at the centre of the exposed hemisphere. (d) A backwash effect takes place and some material flows back to the originally-unexposed hemisphere. (e) A small pile-up occurs at the antipodal region to the centre of the exposed hemisphere.

the chaotic terrain is that the shock waves following the large impact that produced the Caloris Basin, travelling through Mercury by different routes, all met at the antipodal point to the Caloris Basin and interacted there to produce the disturbed surface.

It seems somewhat coincidental that a large impact occurred at one of the sub-solar regions so it is tempting to think that there might be some causal relationship between the position of the Caloris Basin and Mercury's orbital and spin commensurability. In Figure 37.4 we can see that in the process of reforming Mercury after its loss of mantle material there were temporary pile-ups of material at the centre of the damaged region and to a lesser extent in the diametrically opposite region. However, since the whole of Mercury was in a molten, or close to molten, state at this stage of its existence it is hard to see how ripples could have formed and congealed into a permanent form.

There are large lava plains on Mercury, similar in some ways to mare features on the Moon. A feature of Mercury's surface that is not reproduced on the Moon is the presence of long and high scarps, which are wrinkles in the surface due to compression as the planet cooled and shrank. This suggests that the average temperature of Mercury on formation was considerably higher than that of the Moon — compatible with its origin and history as described here.

37.3 The Orbits of Mercury and Mars

Mercury and Mars not only share the distinction of being the two smallest planets in the Solar System but also in having the most eccentric orbits — 0.206 and 0.093, respectively. Their orbital speeds as satellites of Bellona, or possibly Enyo, would have depended on their orbital radii but were probably in the range 20–25 km s^{-1}. Their velocities relative to the Sun would have been a combination of the velocity of the planet relative to the Sun with the velocity of the satellite relative to the planet. With the fairly rapid dispersal of the parent planets, and thereafter only the much smaller residue masses to constrain their motions, the combination of velocities could have given heliocentric orbits with large semi-major axes and eccentricities.

In this respect it should be noted that at 0.4 au from the Sun — the distance suggested for the collision in Section 36.4 — the speed for a circular orbit is 47 km s^{-1} and that to escape from the Solar System is 67 km s^{-1}. The combination of velocities could readily have given resultant speeds towards the upper end of those two values resulting in extended orbits. Such orbits would have taken so much time completely to decay and round-off that the resisting medium might have disappeared before the orbital evolutions were complete.

There may be other explanations for the large orbital eccentricities of the two smaller planets in terms of some other event in the turbulent history of the Solar System, so the above explanation is offered as just one possibility.

Chapter 38

Gods of the Sea and the Nether Regions

The moist star
Upon whose influence Neptune's empire stands...

William Shakespeare (1564–1616), *Cymbeline*

38.1 Neptune and Related Bodies

In Section 7.4 we described the characteristics of dwarf planets and why it was decided to introduce that class of bodies. When Eris was discovered in the Kuiper Belt, and found from the motion of its satellite, Discordia, to be more massive than Pluto, the designation of the latter body as a planet, which had always been controversial, finally came to an end. However, what distinguishes Pluto from the other dwarf planets is its apparent relationship to Neptune. Its projected orbit closely approaches that of Neptune (Figure 7.9) although the two bodies never approach each other closely. The ratio of Pluto's orbital period to that of Neptune is 3:2 on average, sometimes slightly more and sometimes slightly less, and dynamical effects maintain the average and keep the two bodies well apart; in fact Pluto approaches closer to Uranus than it does to Neptune. Again, with an orbital inclination of 17° Pluto is usually well above or well below the plane of Neptune's orbit.

When it comes to physical characteristics, Pluto's mass and radius are in the range of moderate-sized satellites of the Solar System and, having postulated a satellite origin for Mars and Mercury, we might be tempted to consider it as another escaped satellite of one of the colliding planets. We could then explain its orbital relationship with Neptune as due to an interaction with Neptune that swung it from a far-ranging orbit into its present orbit as shown in Figure 38.1. This could have happened at any time, including well after the planetary

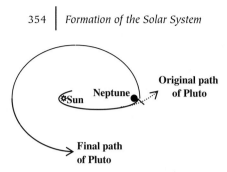

Figure 38.1 Path of Pluto deflected by Neptune (not to scale).

orbits had settled down and the resisting medium had gone, or almost gone, so that the orbit it took up after interacting with Neptune is much the same as it is now. Although there would be little or no medium to give precession of its orbit, so that it became well separated from that of Neptune, the gravitational effects of the other planets could provide the necessary torques to give precession. In addition, there are effects explained by the General Theory of Gravitation that cause precession of an eccentric orbit; for Mercury it amount to 43 seconds of arc per century but for Pluto it is much smaller — about 1° per million years — but it would be significant over the lifetime of the Solar System. The presence of Charon and the other smaller satellites could then be explained as companions it picked up in the hectic environment of the planetary collision and that were carried with it thereafter. Assuming that enough resisting medium was present to give the 3:2 commensurability of the Pluto and Neptune orbits by the process described in Section 30.8, this model explains Pluto's present state and orbit reasonably well — but nothing else.

Neptune itself is a normal planet, structurally similar to Uranus in many ways and with a fairly large axial tilt of 29°; in Section 33.6 an explanation was given for the axial tilts, compositions and large D/H ratios of the ice giants, Uranus and Neptune. However, its family of satellites shows two peculiar features, already referred to in Section 6.5. To reiterate — the largest satellite, Triton, with less than one-third of the mass of the Moon, is in a *retrograde* circular orbit of radius of 350,000 km, less than the Moon's orbital radius (384,000 km). The other odd satellite, Nereid, is of modest size — a mere 340 km in diameter — and in an extremely extended orbit of eccentricity 0.75. The many other small satellites, which were not known before Neptune was visited by a Voyager spacecraft, are all small bodies in circular orbits. One of these, Proteus, is actually slightly larger than Nereid but because it is close to Neptune it was not detected from Earth observation.

38.2 Yet Another Effect of the Collision

Here we will describe an alternative model for explaining the relationship of Pluto to Neptune, that also involves Triton and explains how it came to be a satellite of Neptune in a retrograde orbit. The starting point for this model is as follows. It is proposed that Triton was a one-time satellite of one of the colliding planets that was released into an extended heliocentric orbit taking it beyond the orbit of Neptune. Neptune itself had a family of regular satellites moving in close to circular orbits. The largest of these was Pluto with a mass one-sixth that of the Moon and about 60% that of Triton.

We now consider the position where much of the outer part of the resisting medium has evaporated and has become of low density in the region of Neptune and beyond so that Neptune's orbit is similar to what it is now and decaying very slowly. At the same time Triton is moving in its very eccentric orbit. Since there is still a significant mass of resisting medium in the inner reaches of the Solar System the orbits of Neptune and Triton will continue their differential precessions, as described in Section 33.5, and sometimes the orbits would have intersected, or nearly so. Now we consider a collision between Triton and the satellite Pluto. I investigated this scenario by detailed computer simulation in 1999 and found that, given suitable initial parameters, the results indicate that the system we see today could be the result of such an event.

The starting point is illustrated in Figure 38.2. Triton was in a heliocentric orbit with perihelion 2.6 au and aphelion 55.6 au. Pluto was in a direct circular orbit, of radius 545,000 km, around Neptune.

The details of the collision process and its aftermath can be followed in Figure 38.3. Figure 38.3(a) shows Pluto in its circular orbit being struck by Triton, which was on a path towards the Sun.

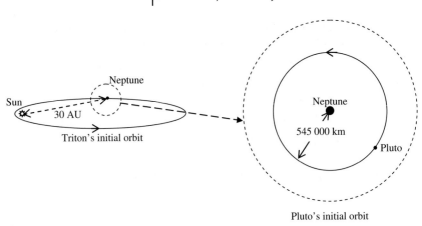

Figure 38.2 The initial orbits of Triton and Pluto before the collision.

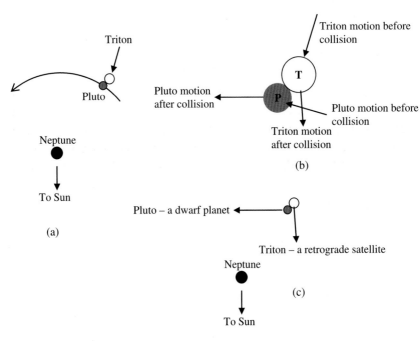

Figure 38.3 The Triton-Pluto collision. (a) Triton, travelling towards the Sun strikes Pluto. (b) Motions before and after the collision. (c) The final outcome.

Figure 38.3(b) shows the motions of both Pluto and Triton before and after the collision. Pluto is struck in such a way that its energy of motion is increased sufficiently to expel it from its orbit around Neptune into a heliocentric orbit, with parameters similar to those at present; the values from the computer simulation were $a = 39.5$ au and $e = 0.253$. By contrast, Triton loses energy of motion and is slowed to the extent that it is removed from its heliocentric orbit and is captured by Neptune into a *retrograde* orbit with semi-major axis 436 000 km and eccentricity 0.88. The final outcome is shown in Figure 38.3(c).

The tidal effect of planets on satellites with retrograde orbits, described by T.B. McCord in 1966, is that they cause the orbits both to round off and to decay, with round off being a fairly fast process. At present Triton's orbit is almost perfectly circular and is slowly decaying. Due to the gradual approach of Triton to Neptune, at some time in the future the satellite will be torn apart by the tidal effects of the planet.

A characteristic of heliocentric orbits, in the absence of some external influence other than the Sun, is that they repeatedly go through the same parts of space. Since Neptune and Pluto were both in the same vicinity at the time of the Triton-Pluto collision it might be expected that the two orbits would continue to have parts in close proximity, although this is not now the situation. While their orbits overlap when seen in projection, in three dimensions they are well separated. The differential precession that brought the orbits of Triton and Neptune together operated similarly to separate the orbits of Pluto and Neptune. This took place until the resisting medium was completely dispersed, at which stage the orbits were stable and separated as they now are. Although the resisting medium had greatly reduced at the time of the collision, it could still have given rise to slow evolution of the orbits of Neptune and Pluto that could have established the 3:2 commensurability, as described in Section 30.8.

One further feature of Pluto can be explained by the Triton-collision hypothesis. Triton gave Pluto a glancing blow that sheared off part of it either in the form of a single coherent large unit or in the form of substantial debris. This is the material that gave rise to Charon and the other smaller satellites. The glancing blow was in a direction that would have spun Pluto into a retrograde direction (Figure 38.3) and also led to the Pluto satellites being in retrograde orbits — as is actually observed. It is interesting to note that a scenario that has been advanced for the formation of the Moon by the American astronomer W. Benz and colleagues, which has received considerable support, involves material being sheared off the Earth by collision with some body with about the mass of Mars — a very similar mechanism to what is suggested here.

Finally we can bring the extreme nature of Nereid's orbit into the total picture. With many satellites of Neptune other than Pluto it is possible that Triton would have passed close enough to one of them greatly to disturb its orbit. Nereid's orbit could then be a manifestation of such a disturbance. Alternatively, and perhaps more probably, Nereid could be a captured body — a large asteroid or small escaped satellite — that underwent a collision in the vicinity of Neptune, lost energy and was captured by the planet into its present highly-eccentric orbit.

Chapter 39

Bits and Pieces — Asteroids, Comets and Dwarf Planets

Gather up the fragments that remain, that nothing be lost.

Gospel according to St John

39.1 The Gap and its Denizens

It was as a consequence of the apparent gap between Mars and Jupiter, revealed by Bode's Law, that a search was begun to find a missing planet, eventually leading to the discovery of Ceres (Section 7.1). Subsequently, large numbers of smaller bodies, now known as asteroids, were discovered mostly, but not always, orbiting between Mars and Jupiter. Some asteroid orbits take them within the orbit of Mars, or even within that of the Earth, and others range well outside the orbit of Jupiter. They are all in direct orbits around the Sun, mostly with eccentricity less than about 0.4 and orbital inclinations less than about 40°, although a few have more extreme orbital characteristics. The majority of the smaller asteroids are of irregular shape. This is illustrated in the NASA spacecraft picture of Ida, an asteroid with dimensions about $56 \times 24 \times 21$ km, which has the peculiar feature that it has a small satellite (Dactyl), seen as a tiny dot on the right-hand-side of Figure 39.1.

The meeting of astronomers that demoted Pluto from planetary status, and set up the category of dwarf planets, also included Ceres in the dwarf planet class of objects. Therefore, any discussion of asteroids should technically exclude Ceres. As it happens, that fits in which the theory of the origin of various kinds of small body that will be described.

Figure 39.1   The asteroid Ida with its satellite, Dactyl.

39.2 Some Ideas on the Origin of Asteroids

Long before images of asteroids were available it was known that they were irregular in shape. Asteroids tumble in space with periods of order two to four hours and, as they do so, their brightness changes according to how much area they present to the Sun's illumination and to the observer. The irregular shapes of asteroids suggested to many planetary scientists that they were the result of some event that broke up a larger body.

An early idea about the origin of asteroids was that they represented the debris from a planet that once existed in the region of the asteroid belt and that had somehow exploded. Various objections were raised to this idea. One is that the total mass of all known asteroids is very small, a small fraction of the mass of the Moon, so it would be necessary to explain where the remaining part of the planet had gone. This could be explained by saying that, over a long period of time, the debris had been swept up by collisions with planets, particularly Jupiter. Another, perhaps more serious, difficulty, raised by the British astronomers W. Napier and R.J. Dodd in 1973, is that there is no known source of energy that could cause a planet spontaneously to explode and completely disintegrate.

An alternative explanation that has been advanced for asteroids, related to the solar-nebula model for planetary formation, is that asteroids are planetesimals (Section 21.3) that were left over after planet formation was complete. However, there are also difficulties with this idea. Firstly, the physical composition of some meteorites shows that at some time their material has been molten. Secondly, meteorites consist of two main types — irons and stones — reflecting the type of material that comprises them, with comparatively few containing substantial proportions of both iron and stone. Both these pieces of evidence suggest that asteroids were derived from larger bodies that were molten and within which a substantial gravitational field existed that separated material into layers of different densities.

If material is brought together to form a large body then gravitational energy is released in the form of heat (Section 36.2). Theory shows that if a spherical body of mass M is assembled by bringing material together from a large distance then the heat energy released is

$$H = C\rho M^{5/3}, \tag{39.1}$$

where ρ is the density of the material and C is a constant. Thus, the amount of heat generated per unit mass of material is

$$H/M = C\rho M^{2/3}, \tag{39.2}$$

which increases with increasing M. For this reason the accumulation of a sufficiently massive body can give melting of its material. It can be shown that, for stony material, the critical radius is about 1,200 km, much greater than that of Ceres but smaller than that of the Moon. We have already noted that the outer regions of the Moon were molten when it was first formed, so giving the volcanism that was discussed in Section 36.3. Within a body of a size that would give melting, material would segregate through gravity to give an iron core and a stony mantle, so giving the separation of material seen in meteorites and inferred in asteroids. Since bodies the size of asteroids could not melt and give segregation of iron and silicate, this gave rise to the idea of *parent bodies*, produced by small accumulations of planetesimals, the collisions and disruptions of which produced asteroids. It is suggested that parent bodies, smaller than the critical mass for melting through the release of gravitational energy, could have melted due to the presence of a radioactive isotope, aluminium-26, when they first formed; the parent bodies would still be large enough to give gravitational fields that would, given enough time, separate iron and silicates. Aluminium-26 has a short half-life of about 720,000 years, which means that it produced all its heating effect within a few million years and that it has now completely disappeared. Actually, if aluminium-26 were around in the early Solar System in the quantities suggested then even small asteroids, a few kilometres in diameter, could have melted.

Much of what we know about asteroids comes from a study of meteorites, fragments of asteroids that land on Earth. A comparison of the reflection spectra from asteroids, determined by telescopic observations, with those of various types of meteorites, determined from laboratory experiments, enable the composition of various types of asteroid to be inferred. Figure 39.2 shows a comparison of reflection spectra for various pairs of asteroid and meteorite.

There are three main types of asteroid: C-type, similar to carbonaceous chondrites (Section 7.3.1); S-type corresponding to other stony

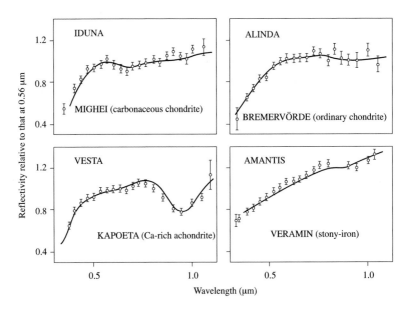

Figure 39.2 Reflection spectra of asteroids (points with error bars) and of matching meteorites (full lines).

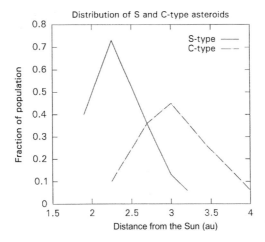

Figure 39.3 The distribution of distances from the Sun of C and S-type asteroids.

meteorites; and M-type, related to iron meteorites (Section 7.3.2). The C- and S-types account for about 80% of all asteroids and their distribution in terms of their distance from the Sun is shown in Figure 39.3. Although the distributions overlap it is quite clear that, on average, the C-type asteroids occupy a region further from the Sun.

39.3 Comets and the Kuiper Belt — General Considerations

In Section 7.3.2 different categories of comets were described according to the kinds of orbits they occupied. The crudest categories were *short period*, with periods less than about 200 years that meant they spent most time in the planetary region, and others designated as long period. A characteristic of short period comets is their limited lifetime; after about 1,000 perihelion passages they lose their volatile content and cease to be seen as comets. They must therefore be replenished from some source and that is believed to be the Kuiper Belt (Section 7.4). At the other extreme there are those long period comets that originate in the Oort Cloud. One idea that has been suggested is that the Oort Cloud was formed from bodies captured by the Solar System when it passed through a dark cool molecular cloud, which could possibly contain comet-like bodies. Some long-period comets are the so-called *new comets* that originate in the Oort Cloud, at distances of tens of thousands of au from the Sun. These comets are new in the sense that they have never been seen before (their periods can be millions of years) and they are unlikely to appear again with a similar orbit. The reason for this is that, coming from such a large distance, these comets are very loosely bound to the Solar System. When they enter the planetary region, perturbation by the planets either adds or subtracts energy of motion. The likely amount of energy provided is such that if it is added they will almost certainly leave the Solar System but if is subtracted then the comets will move into a much more tightly bound orbit, with a much smaller aphelion distance — very different from that with which it first approached the Sun.

So the picture that we now have for cometary bodies is that they exist in the Kuiper Belt, just outside the planetary region, and they also exist at enormous distances from the Sun, out to almost half way to the nearest star. If we combine this information with that from asteroids then we see that the general pattern is that the distance of bodies from the Sun is related to their volatile content. The S-type asteroids with the least volatile content are closest to the Sun, then the C-type containing volatiles are further out and finally comets with a high, even dominant, volatile content are further away still. Of course, the idea that a body should have greater volatile content the further it is from the Sun, which is a considerable source of heating, seems quite natural but we now look at another explanation in terms of the planetary collision.

39.4 The Planetary Collision Again!

Figure 35.2, which follows the progress of the planetary collision, shows that the planetary material is thrown out in all directions. The material that is of interest is the solid debris coming from the core, mantle and ice regions. It might be expected that outer material, with little outside it to hinder its loss, would be thrown out to the greatest distance and that the innermost material would end up closest in.

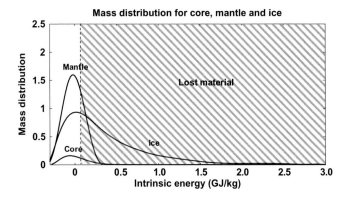

Figure 39.4 The distribution of dispersed material from the collision. The graphs show the distribution of mass in Earth units per unit intrinsic energy (GJ[1]/kg).

Analysis of the destination of the ejected material shows that this is so. About two-thirds of the ice, one-third of the mantle material and about one-third of the core material is expelled from the Solar System. This is shown in Figure 39.4, expressed in terms of the energy per unit mass (specific energy) of the ejected material. If this is negative then the negative potential energy due to the Sun's attraction has a greater magnitude than the positive kinetic energy due to motion and so the body is retained. Conversely, if the intrinsic energy is positive then the energy of motion can overcome the Sun's attraction and the body is lost from the Solar System.

Another point of interest is the inclinations of the retained material, shown plotted against eccentricity in Figure 39.5. Any inclination below 90° corresponds to a direct orbit and it will be seen that virtually all the core and mantle material is in direct orbits, which agrees with what is observed for asteroids. However, more of the ice debris, which leaves the region of the collision with higher velocities and would become the source of comets, is in retrograde orbits.

All the material retained in the Solar System will initially be in orbits that have perihelia not too distant from the collision region so

---

[1] 1 GJ (gigajoule) = 10^9 J.

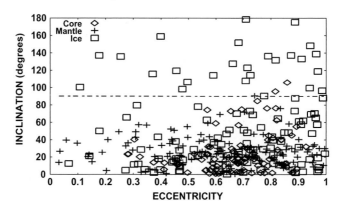

Figure 39.5   The eccentricities and inclinations of the retained material.

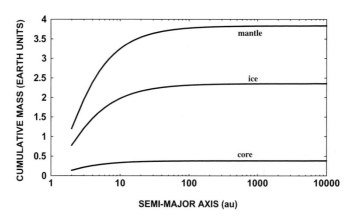

Figure 39.6   The cumulative distribution of mass for core, mantle and ice material for different maximum semi-major axes.

that different orbits can be approximately categorized by their semi-major axes. Figure 39.6, which shows the information in Figure 39.4 in a different form, gives the cumulative mass out to different values of semi-major axis so that, for example, the total mass of mantle material with semi-major axes less than 10 au is 3.24 $M_\oplus$.

We now consider explanations for the various classes of smaller bodies in relation to the collision model and the initial orbits of the debris.

39.5 Asteroids

Asteroids originated from core and inner mantle material that ranged less far out from the region of the collision. The initial orbits would all return repeatedly to the region of the collision and, as can be deduced from Figure 39.6, much of it would have had aphelia in the region that would eventually be occupied by the present major planets, and so would not have survived. Again, there would have been many interactions of debris fragments in the inner Solar System, both with the Earth and Venus as they settled down into their final orbits, and with each other. The asteroids that exist today are the few that ended up in orbits, such as in the asteroid belt or in the region between Saturn and Uranus, now occupied by Charon, and probably other bodies, which would enable them to survive once the Solar System had reached its present configuration. This would have been a tiny fraction of the original material; the total mass of all asteroids is estimated to be about 3×10^{21} kg, or 4% of the mass of the Moon.

39.6 Comets and the Oort Cloud

An argument that has been advanced to support an external origin for Oort Cloud comets is that, since their D/H ratios can be up to 100 times the general Universal value, they could not have a solar-system origin. The model of colliding planets containing ice with a high D/H ratio weakens this argument.

In 1983 the British astronomer, Mark Bailey, proposed that there is an inner reservoir of comets which are drawn outwards to replenish the Oort Cloud when it is depleted by a very severe disturbance, such as when a star passes near or through it or when the Solar System passes through a Giant Molecular Cloud. Certainly, there is no indication from Figures 39.4 and 39.6 that there was any break in the distribution ranging from the inner Solar System out to Oort-Cloud distances but the way that orbits have been modified must be considered before it can be said that this model of comet formation supports the Bailey hypothesis. If the continuity of the distribution has persisted throughout the period of periodic disruption of the Oort Cloud and its reinforcement from within, then we must interpret the observed parts of the Kuiper Belt as the inner boundary of this distribution. As previously stated, short-period comets are those coming from the Kuiper Belt perturbed by Neptune and *new comets* are those from the outer part of the distribution, the Oort Cloud, which are perturbed by sources exterior to the Solar System. Since there are no major sources of perturbation for the great majority of comets, those neither close in nor far out, there is no way in which their presence can be detected.

At the time of the collision the other major planets were still on evolving orbits with aphelia probably of the order of hundreds of au and, in the nature of eccentric orbits, for most of the orbital period they would be in the aphelia regions. This gives rise to the possibility of interactions between wide-ranging debris and a major planet similar in form to that shown in Figure 38.1, where the cometary debris would play the role of Pluto and Neptune would be a general major planet. This would give cometary orbits with perihelia well outside the present region of the planets, so that if they could survive until the orbits of the major planets had fully evolved they would survive indefinitely thereafter.

This process is most efficient if the planet and debris are on orbits with similar aphelia but it is also necessary for the bodies to approach each other closely, which requires them to be not too far from coplanar. For this reason, although some debris *can* be moved in this way to orbits keeping them clear of the inner Solar System, it is unlikely to give an explanation to move much of the debris out of harms way.

Another kind of interaction involves the effect of the medium on the newly formed cometary bodies. For bodies the size of comets, of kilometre dimensions, the main resistance is due to the ram pressure they experience due to the impact of the medium on the comet. For a spherical comet the force experienced will be

$$F = \pi \rho a^2 V^2 \tag{39.1}$$

where ρ is the local density of the medium, a the radius of the comet and V the velocity of the medium relative to the comet; the force on the comet is in the direction of *V*. The effect of such a force has been found for a comet of mass 7×10^{12} kg, of density 500 kg m^{-3} (published estimates are between 100 and 1,000 kg m^{-3} — they are very loose structures), with original perihelion 0.5 au and various original semi-major axes. The medium had a total mass of 40 M_J with a distribution of density similar to that seen in Figure 30.7 and splayed out perpendicular to the mean plane of the medium as seen in Figure 30.3. The results are shown in Figure 39.7.

It will be seen from the figure that all orbits with original semi-major axes greater than about 60 au will end up with perihelia beyond the orbit of Neptune and within the Kuiper Belt region. Their aphelia stretch out to several hundred and even thousands of au and represent the inner cloud of comets postulated by Bailey as a reservoir for the replenishment of the Oort Cloud. It should be noted that the orbits of asteroids, with their much higher densities, would be much less affected by this kind of process, although it might move some of them into the safety of the asteroid-belt region.

When Jan Oort first described the structure of what became known as the Oort Cloud he made the assumption that, due to perturbation in random directions by passing stars and other influences over a long period of time, in any region of the cloud the velocities were uniformly distributed in all directions and the ends of the

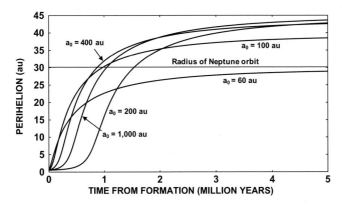

Figure 39.7 The changes of the perihelion with time for comets with original perihelia 0.5 au and various initial semi-major axes, a_0.

vectors representing them uniformly filled a sphere with the radius of the escape velocity for the region. From the proportion of the corresponding orbits that would produce visible comets, and the frequency of the observations of new comets, those coming from the Oort Cloud, he estimated that there were 2×10^{11} comets in the cloud. From this, a variety of estimates have been made about the total mass of the cloud and they range from about one to several Earth masses. It would be difficult to explain such a mass from an analysis of the outcome of the planetary collision hypothesis. From Figure 39.6, the mass of icy material with semi-major axis greater than 60 au, and which may end up within the Kuiper Belt and beyond, is about $0.07\ M_{\oplus}$ or 4×10^{23} kg.

The basis of Oort's assumption about the velocity distribution of comets in the cloud reflected the general view at the time that the cloud was formed when the Solar System originated, with comets either coming from outside the Solar System or somehow being formed at large distances from the Sun. With this assumption many comets are in orbits of moderate eccentricity with perihelia tens of thousands of au and many others on orbits ranging out to distances approaching those of distant stars. On the other hand, the model given by Bailey, of the existence of an inner reservoir of comets that reinforce the Oort Cloud when it is disrupted, gives a completely different scenario. Comets perturbed from the outer parts of the inner reservoir to orbits ranging out to tens of thousands of au will be on highly elliptical paths with smaller perihelia and the probability per comet that they could give visible new comets is much higher than the Oort model would suggest. Again, from dating craters on the Moon it has been deduced that about 500 million years ago there was a great increase of projectiles in the inner Solar System. This suggests that a major perturbation of the Oort Cloud may have occurred then and, if reinforcements to the cloud had arrived from the inner reservoir at the same time, then larger eccentricities, and hence smaller perihelia, of Oort-Cloud comets would be expected at the present time.

A new estimate has been made on the basis of considering Oort Cloud comets with aphelia out to a distance of 100,000 au, almost

40% of the distance to the nearest star, with the assumption that the allowed velocities are those with magnitudes less than the escape speed that also give a perihelion less than 5,000 au. On the basis of this model the expected number of comets in the Oort cloud is 2.5×10^9. This number could have been larger in the past but, even so, the expected number of comets in the original Oort Cloud would be far less than Oort's estimate. Given, say, the existence of 10^{10} comets in the early Solar System, plus the deduced mass of material of 4×10^{23} kg, would give a mean mass per comet of 4×10^{13} kg; this would be the mass of a spherical comet of radius 2.5 km and density 600 kg m^{-3}, a very reasonable number.

39.7 Dwarf Planets

The two large colliding planets would have had many large satellites among which we have identified the Moon, Mars and Mercury. There are five dwarf planets at present — Ceres, Pluto, Eris, Makemake and Haumea. Their masses are all within the range observed for the present satellites of the Solar System and, with the exception of Pluto, we identify these as satellites of the colliding planets that were diverted into orbits within the Kuiper Belt. In this respect it should be noted that satellites of the colliding planets would be released into orbits close to the plane of the planetary orbits so that interactions like that in Figure 38.1 would have a higher probability of occurring. Again, the annular form of the resisting medium, as shown in Figure 30.7 leads to an increase of the perihelion; those satellites perturbed into orbits with a perihelion within the Kuiper Belt would survive and others would eventually be swept up by a major planet. The exception is Ceres, left in an orbit within the asteroid belt.

There would have been many other satellites, possibly some more massive than the present dwarf planets. Many of these may have escaped from the Solar System and others absorbed by major planets. Others, yet to be discovered, might be in outer regions of the Kuiper Belt. Another possibility is that one or more of them may have reached the Oort Cloud. It has sometimes been suggested that the tendency of some new comets to come from similar directions with similar orbital parameters may be due to the presence of major perturbing bodies within the Oort Cloud itself.

39.8 Features of Meteorites

In the 1960s, it was realised that the compositions of meteorites could be explained in terms of the way that minerals would condense from a hot silicate vapour and this revived an interest in a nebula theory for the origin of the Solar System (Section 19.1). In particular chondrules — small, glassy spheres — are exactly what would be expected from a cooling silicate vapour that would produce molten droplets in the form of spheres due to the action of surface tension, which would then freeze and become solid. Another feature of chondrules, that the minerals within them are unequilibrated (Section 7.3.1), indicates that they cooled very quickly; the vaporized material thrown outwards by the colliding planets would, indeed, have cooled quickly. Although a planetary collision may be thought of as a massive event from a human perspective, from an astronomical point of view it is quite small scale and the ejection of material and its subsequent cooling would have taken place on a timescale of hours, or even minutes.

39.9 A Summary of the Production of Small Bodies

In a certain sense the collision hypothesis explains the formation of small bodies very efficiently. All the expected product of the collision — core material, mantle material, icy material, quickly-cooling chondrules and satellites of the colliding planets — are evident in the present structure of the Solar System and all small-body features of the Solar System can be explained in terms of the collision. There are some very tiny bodies we have not mentioned. Small collections of dust particles, called *Brownlee particles*, are almost certainly the dust debris from comets. There are also *Interplanetary Dust Particles* of typical size a few hundred microns that show high D/H ratios of up to 0.01. These are often described as the product of distant stars that have drifted through interstellar space and entered the Solar System but, alternatively they can represent the dusty debris from high D/H icy material in the colliding planets that was thrown clear and was not involved in the nuclear explosion.

A final point: on the basis of the planetary-collision theory, if dwarf planets were redefined as ex-satellites of the colliding planets that had gone into independent heliocentric orbits and were massive enough to take up a spherical form then there would be six of them — Ceres, Eris, Makemake, Haumea, Mercury and Mars — and once again poor Pluto would be out on its own!

Chapter 40

Making Atoms with a Biggish Bang

I can trace my ancestry back to a protoplasmal
primordial atomic globule.

W.S. Gilbert (1836–1911)

40.1 Let's Find Out More About Isotopes

Sometimes in a newspaper or magazine article the word 'isotope' turns up — often in the context of something that is made in a nuclear reactor and has some medical use. Well then — what is an isotope? In Figure 26.1 we showed a representation of a carbon atom, the one labelled carbon-12 in Figure 40.1. The nucleus contains six protons (red), each with a positive charge and six uncharged neutrons (black). There are also six negatively charged electrons (blue) in the region around the nucleus to balance out the proton charges and to make the atom as a whole electrically neutral.

Electrons have a very tiny mass that can be disregarded in respect to the total mass of the atom. Protons and neutrons have masses that are nearly equal and 1,840 times greater than that of the electron. The '12' in the designation 'carbon-12' means that the atom has 12 units of mass. We do not have to specify the number of protons (equal to the number of electrons) because it is six protons in the nucleus that makes the atom carbon and not some other kind of atom. All living material contains carbon, and that includes us. If we could examine the individual carbon atoms in our bodies *most* of them would be carbon-12. However, just over 1% of them would be different — we would find that the nucleus contains seven neutrons, not six (Figure 40.1). Is it still carbon? Yes it is — it is a perfectly normal carbon atom (the six protons ensure that) and it will behave

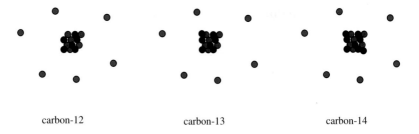

| carbon-12 | carbon-13 | carbon-14 |

Figure 40.1   The structures of carbon-12, carbon-13 and carbon-14.

just like any other carbon atom in a chemical reaction. It is an *isotope* of carbon called carbon-13 because the atom has 13 units of mass. Just over 1% of all the carbon atoms on Earth, from whatever source, are carbon-13.

Actually a very diligent search amongst all these carbon atoms would find another isotope, carbon-14, that has eight neutrons in its nucleus. It is, however, a somewhat uncomfortable atom and would rather be something else. Carbon-14 is *radioactive* and once in a while, apparently at random, a carbon-14 atom undergoes a transition that turns it into a nitrogen atom. If there is a large number of carbon-14 atoms then in 5,730 years one half of them will have transformed into nitrogen; we say that the *half-life* of carbon-14 is 5,730 years. Despite its short half-life compared with the age of the Solar System, carbon-14 occurs naturally since it is continuously produced by the action of cosmic rays on nitrogen-14 in the upper layers of the atmosphere. All living matter contains carbon-14 and archaeologists use it to date specimens of dead organic matter. It must be stressed again that all the carbon isotopes are *chemically* identical and it is only in their masses that they differ. Other carbon isotopes are possible, e.g. carbon-11, and can be made in nuclear reactors but they are radioactive with a short half-life and therefore have no permanent, or even long term, existence.

There is a very convenient shorthand and informative way of writing carbon-12, which is $^{12}_{6}C$. The C is the *chemical symbol* that represents carbon. The numbers, 12 and 6, give the mass of the atom (number of protons + neutrons) and the number of protons respectively.

Actually, the six is redundant and scientists, who know how many protons there are in a carbon atom, will simply write ^{12}C. However, we shall continue to use the redundant form. These, then, are the isotopes of carbon we have considered so far —

$$^{12}_{6}C \qquad\qquad ^{13}_{6}C \qquad\qquad ^{14}_{6}C$$

Here is a summary of what we have learnt above. For many atoms there is more than one stable isotope. There may also be naturally occurring radioactive isotopes, which change into some other kind of atom, with a particular half-life. There is one other fact that we should know. When we examine carbon on Earth, from any source, we find the ratio $^{12}_{6}C : ^{13}_{6}C = 89.9:1$. If we find an object with a significantly different ratio of carbon isotopes then we may reasonably infer that it is not of terrestrial origin. A similar statement can be made about the isotopic ratios of other atoms. For many atoms there is a terrestrial isotopic composition and any specimen giving a different composition is probably of non-terrestrial origin.

40.2 Isotopes in Meteorites

We know that meteorites have not come from Earth but are mostly bits knocked off asteroids. Nevertheless, while the model that we have for producing planets would not necessarily make them all the same chemically, we might expect them all to be the same in terms of their isotopic composition. If we found a meteorite with a $^{12}_{6}C : ^{13}_{6}C$ ratio very different from 89.9:1 then we might wonder what had happened to its material or, indeed, whether it was really from the same original source as terrestrial material. It is not easy significantly to change isotopic composition. Small changes can occur, due to heating effects or other physical conditions, but these changes would be both small and explicable.

Actually, meteorites show very marked differences of isotopic composition from terrestrial material for many different kinds of atom. Here we shall just deal with a selection of such differences, which are usually designated as *isotopic anomalies*.

40.2.1 *The carbon anomaly*

In some chondritic meteorites (Section 7.3.1) there exist grains of the mineral silicon carbide, chemically represented as SiC. We already know that C is the chemical symbol for carbon and Si is the symbol for silicon. The chemical representation shows us that one atom of silicon combines with one atom of carbon to form a silicon carbide unit. When the ratio $^{12}_{6}C : {}^{13}_{6}C$ is found for the carbon in samples of SiC from different meteorites they are found to vary greatly and many are much less than 89.9 — down to 20, or even less. There is more of the heavier isotope $^{13}_{6}C$ than is found on Earth and this isotopic anomaly is referred to as *heavy carbon*.

An explanation that has been given for heavy carbon is that it is the result of an influx of material into the Solar System from distance carbon stars. These are stars that are known, from analysis of the light they emit, to be rich in carbon — and the carbon they contain is heavy carbon. It has been suggested that the various isotopic compositions of SiC grains can be explained if they originated in six or more carbon stars. The grains then drifted across interstellar space and were incorporated into meteorites.

40.2.2 *The nitrogen anomaly*

The anomalies observed for nitrogen are also based on measurements on silicon carbide grains. The nitrogen gets trapped in small interstices within the grains and is released by heating. Nitrogen has two stable isotopes, $^{14}_{7}N$ and $^{15}_{7}N$, and in the terrestrial atmosphere the ratio $^{14}_{7}N/^{15}_{7}N$ is 270. Most of the nitrogen from silicon carbide is *light nitrogen* with a ratio of up to 2,000 but, curiously, *heavy nitrogen* also occurs with a ratio as low as 50.

40.2.3 *The oxygen anomaly*

The normal oxygen (chemical symbol O) that we breathe consists of three isotopes represented as $^{16}_{8}O$, $^{17}_{8}O$ and $^{18}_{8}O$. You should now be in a position to know how many protons, neutrons and electrons those isotopes contain. The standard terrestrial ratios for those isotopes are

$$^{16}_{8}O : {}^{17}_{8}O : {}^{18}_{8}O = 0.953 : 0.007 : 0.040$$

and this mixture is known as SMOW (Standard Mean Ocean Water). Given that various thermal processes can change those ratios a little, but in a predictable way, oxygen specimens from the Earth and Moon seem quite compatible.

Samples of oxygen from some ordinary chondrites and carbonaceous chondrites give oxygen isotopic ratios that cannot be explained by simple physical processing of normal terrestrial oxygen. However, they *can* be interpreted as being admixtures of SMOW with different amounts of pure $^{16}_{8}O$, but the problem is to find some source of pure, or nearly pure, $^{16}_{8}O$. One explanation that has been offered is that it is produced by the action of alpha particles (the nuclei of helium atoms, represented by $^{4}_{2}He$) on $^{12}_{6}C$ in distant stars. This pure $^{16}_{8}O$ is then incorporated into dust grains that travel to the Solar System and become incorporated in meteorites. Subsequently, normal solar-system oxygen infiltrated the grains displacing most, but not all, of the $^{16}_{8}O$ so giving a final content that is rich in that oxygen isotope.

40.2.4 *The magnesium anomaly*

There are three stable isotopes of magnesium (chemical symbol Mg) in the ratios

$$^{24}_{12}Mg : {}^{25}_{12}Mg : {}^{26}_{12}Mg = 0.79 : 0.10 : 0.11.$$

The white high-temperature inclusions in carbonaceous chondrites (Section 7.3.1) were found to contain excess $^{26}_{12}Mg$ in proportion to the amount of aluminium in the sample. There is only one stable isotope of aluminium, $^{27}_{13}Al$, but, as mentioned in relation to the possible melting of asteroids (Section 39.2), there is the radioactive isotope of aluminium, $^{26}_{13}Al$, with a half-life of 720,000 years. The product of its decay is $^{26}_{12}Mg$.

The explanation given for the magnesium anomaly is that when the meteorite formed the aluminium within it contained a small fractional component of $^{26}_{13}$Al — of the order of one part in from 10,000 to 100,000. This then decayed leaving an excess of $^{26}_{12}$Mg in the meteorite. Different mineral grains in a meteorite would have different chemical compositions but the aluminium in them will all be derived from the same source and hence have the same ratio of $^{26}_{13}$Al : $^{27}_{13}$Al. The excess of $^{26}_{12}$Mg in a particular mineral grain within the meteorite would then be proportional to the amount of aluminium (hence also of $^{26}_{13}$Al) that it contained.

In Section 26.4 we mentioned that the event that triggered the formation of a cool dense cloud was a supernova and this would have produced a wide range of isotopic products, including $^{26}_{13}$Al. However, it takes about 60 million years for the cool dense cloud to be produced — more than 80 half-life periods for $^{26}_{13}$Al — by which time all the $^{26}_{13}$Al would have disappeared. To explain how $^{26}_{13}$Al could have been present in the early Solar System it has been suggested that just before the Solar System formed there was a second supernova somewhere in its vicinity.

40.2.5 *The neon anomaly*

Neon (chemical symbol Ne) forms a small proportion of the air we breathe and it is best known as the material that gives rise to the orange glow of 'neon signs'. It is chemically inert, which means that it does not normally produce chemical compounds, so when it occurs in meteorites it is in the form of individual neon atoms. These are trapped in atomic-size cavities but they can be released by heating the meteorite. In fact, if neon or other gases are found in meteorites it is certain that these meteorites were not substantially heated after the neon was incorporated — otherwise the neon would have escaped.

Normal terrestrial neon has the composition

$$^{20}_{10}\text{Ne} : {}^{21}_{10}\text{Ne} : {}^{22}_{10}\text{Ne} = 0.905 : 0.003 : 0.092.$$

Specimens of neon from different solar-system sources are very variable in isotopic composition and it has been deduced that there may be three separate sources and that the neon we normally measure in meteorites is formed from different admixtures of these sources.

There is another kind of neon in meteorites that cannot be explained as a mixture of three basic components. Some meteorites contain pure, or almost pure, neon-22, which is a 9% component of terrestrial neon. This anomalous neon is called neon-E. There is no conceivable natural mechanism for completely separating neon-22 from a mixture of isotopes so some other explanation is required. The most plausible source of this neon is by the decay of a radioactive isotope of sodium (chemical symbol Na), $^{22}_{11}$Na. The only stable isotope of sodium is $^{23}_{11}$Na. One scenario that has been suggested is that $^{22}_{11}$Na was produced in a supernova and then was incorporated, together with stable sodium, in minerals. The $^{22}_{11}$Na decayed and the resultant $^{22}_{10}$Ne was trapped within the mineral grain. The only difficulty with this scenario is that $^{22}_{11}$Na has a very short half-life — 2.6 years. This means that the radioactive sodium has to be produced in a supernova and then incorporated into a *cool* solid body within a period of less than 20 years. If the body were not cool at the time of its incorporation then the neon would not have been retained.

40.2.6 *Hydrogen isotopes*

In Section 35.2 we discussed the two stable hydrogen isotopes and discovered why it is that deuterium is heavily concentrated in the ices of cool dark clouds, protostars and, presumably, early protoplanets. Spacecraft have been able to collect micron-sized particles, called interplanetary dust particles (IDPs), from the region above the stratosphere, which prove to be rich in deuterium with D/H up to 0.01. We have already given a possible explanation of IDPs, that they represent micron-size mineral grains originally trapped in the otherwise pure ices (mostly water, methane and ammonia) that formed the outermost solid layer of one or both of the colliding planets. Hydrated minerals, for example serpentine that is highly hydrated and is an important component of some carbonaceous chondrites, would also exhibit the D/H ratio of the ices within which they existed.

40.3 For the Last Time — The Outcome of a Planetary Collision

In Section 35.2, when the composition of the colliding planets was described, the ice layer was taken as having a D/H ratio of 0.01, based on observations of the ratios in icy grains and protostars in star-forming clouds. The temperature of the ice layers reached a triggering temperature for D-D reactions of about 3×10^6 K, which further raised the temperature to more than 10^8 K at which temperature other reactions could take place, leading to a massive nuclear explosion.

In 1995, Paul Holden and I investigated what would be the outcome of the nuclear explosion in terms of the isotopes that would be produced by the reactions. The program considered 548 different nuclear reactions involving the most common elements that would be present in a silicate-ice mixture. We also had to include some radioactive decays in the program since some of the products of the reactions are radioactive and what we observe now is the product of their decay. At the time there was no information available about the expected D/H ratios in protostars, and the planets produced from them, so Holden and I took the value for Venus, 0.016. In the light of more recent information in 2011, I repeated the calculations with a ratio of 0.01; there were differences in the detailed results but the general pattern was unchanged.

The outcome of this calculation was able to explain all the isotopic anomalies described in this chapter, plus others we have not dealt with here.

Carbon

The reactions produced $^{13}_{6}C$ in some quantity and the formation of $^{13}_{7}N$, which decays to $^{13}_{6}C$ with a half-life of 9.97 minutes, also had to be taken into account. The full range of heavy carbon compositions could be explained in terms of the reaction products.

<u>Nitrogen</u>

The amount of original $^{14}_{7}\text{N}$ slightly reduces during the progress of the nuclear explosion but the amount is augmented by two other sources. One is the production of $^{14}_{8}\text{O}$ that decays to $^{14}_{7}\text{N}$ with a half-life of 70.6 seconds and the other is the production of $^{14}_{6}\text{C}$ that decays to $^{14}_{7}\text{N}$ with a half-life of 5,730 years. The $^{14}_{7}\text{N}$ produced from $^{14}_{8}\text{O}$ will be trapped as a gas in SiC grains from the beginning of their formation but the $^{14}_{6}\text{C}$ will first be part of the solid grain and the $^{14}_{7}\text{N}$ will be released into the interstices to augment what is already there as it decays. The combination of the original $^{14}_{8}\text{O}$ contribution plus that of $^{14}_{6}\text{C}$ explains the formation of 'light nitrogen'.

There is prolific production of $^{15}_{7}\text{N}$ towards the end of the nuclear explosion when the temperature is highest but some $^{15}_{7}\text{N}$ is also produced by the decay of $^{15}_{7}\text{O}$ with a half-life of 122.2 seconds. It is at this stage of the nuclear reactions that the occasional samples of 'heavy nitrogen' are produced. The formation of various carbon and nitrogen isotopes is shown in Figure 40.2.

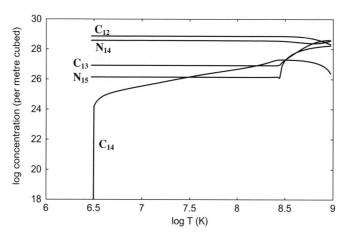

Figure 40.2 The concentrations of isotopes of carbon and nitrogen as the temperature within the nuclear reaction region increases.

Oxygen

In the heart of the explosion the reactions that took place virtually destroyed all the $^{17}_{8}O$ and $^{18}_{8}O$ but left the $^{16}_{8}O$ intact. There was some $^{17}_{8}O$ produced by the decay of radioactive fluorine, $^{17}_{9}F$ with a half-life of 1.075 minutes and of $^{18}_{8}O$ by the decay of $^{18}_{9}F$ with a half-life of 1.83 hours. Figure 40.3 shows the amounts of the three stable isotopes of oxygen as the temperature changed, with the fluorine contributions added into the amounts of $^{17}_{8}O$ and $^{18}_{8}O$. The high-temperature product, when there is a great reduction in the amounts of $^{17}_{8}O$ and $^{18}_{8}O$, provided the source of almost pure $^{16}_{8}O$ that mixed with other normal oxygen to give the anomalous oxygen.

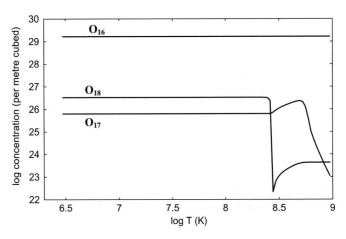

Figure 40.3   The variation of the isotopes of oxygen (including radioactive fluorine) with temperature.

Magnesium

The magnesium anomaly, evident as an excess of $^{26}_{12}Mg$ is due to the copious formation of $^{26}_{13}Al$ at temperatures above 2.5×10^8 K. In considering the isotopic composition of magnesium, we must take account of the formation of unstable $^{24}_{11}Na$ and $^{24}_{13}Al$, which decay to $^{24}_{12}Mg$ with half-lives of 15 hours and 2.1 seconds respectively and of $^{25}_{11}Na$ and $^{25}_{13}Al$, which decay to $^{25}_{12}Mg$ with half-lives 59 and 7.1 seconds respectively. Including these components the variation in the magnesium isotopes, and also of $^{26}_{13}Al$ and $^{27}_{13}Al$, with temperature are shown in Figure 40.4.

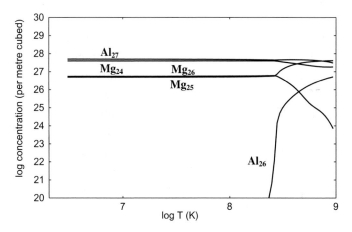

Figure 40.4 The production of magnesium and aluminium isotopes during the nuclear explosion.

There are some observations that show a *deficit* of $^{26}_{12}$Mg and this can be related to the steep fall-off in the amount of that isotope at high temperatures.

Neon

A sufficient quantity of $^{22}_{11}$Na was produced to explain the production of neon-E. We noted that the idea that $^{22}_{11}$Na was produced in a supernova and was then incorporated in cold material on a timescale less than 20 years. This is plausible for production in a planetary collision. The scale of a planetary collision is small by astronomical standards and it can be shown that expansion, the formation of grains and cooling would all take place within hours or days.

The other mixtures of neon isotopes, which had previously been explained as mixtures from three different neon sources, were also explained as products of the nuclear reactions.

40.4 Deuterium in the Colliding Planets and Other Bodies

The production of isotopic anomalies depends on the reactions that took place in a nuclear explosion. The nuclear explosion involving elements heavier than deuterium depended on reaching a high-enough temperature to initiate it and that temperature depended on the amount of deuterium available. The minimum D/H ratio required within an icy environment to set off other nuclear reactions is about 0.005, e.g. about one-half of that indicated by IDPs.

It is proposed that comets and asteroids (and hence meteorites) were derived from those parts of the core and mantle regions of the colliding planets that were remote from the exploding region and hence were unaffected, or only partially affected, by the high temperatures that were generated. These bodies show evidence of enhanced D/H ratios. Some material would have been mostly ice containing very small quantities of silicate. When the ice evaporated there would have been a residue of silicate fragments, including fine dust that would be in the form of IDPs. The deuterium bound into hydrated minerals in IDPs would be more resistant to exchange with external hydrogen than would be that in volatile molecules so, as expected, IDPs show a higher D/H ratio than do comets.

Chapter 41

Is the Capture Theory True?

*If God were to hold out enclosed in His right hand all Truth, and
in His left hand just the active search for truth, although with the
condition that I should ever err within, and should say to me:
Choose! I should humbly take his left hand and say: Father! Give
me this one; absolute truth belongs to Thee alone.*

G.E. Lessing (1729–81)

The answer to the question that forms the subject of this chapter was
given in Chapter 1 — no theory can ever be designated as true. This
is not just being evasive, like not answering the question 'Will you
beat your wife again?' with a simple 'yes' or 'no'. In this case the
question is meaningless because it is impossible in principle to know
what the answer is. If the question is changed to 'Is the theory plausible in the light of present knowledge?' then there is the possibility
of giving a definitive answer.

None of the mechanisms that have been described, from the initial cooling of the interstellar medium to the tidal and impact interactions that gave various features of the planets, is outside the
boundaries of what science will comfortably allow. In addition, the
probabilities of the various events have all been shown to be accept-
ably high. One has to be a little cautious with probability arguments.
As human beings we are all individually highly improbable. The probability that the particular combination of parental genes was formed
that produced a particular individual is incredibly low. However,
given the statistics of reproduction, someone has to be born. For the
model described here for the initial planetary orbits, the probability
that Bellona and Enyo collided is quite small — let us say 0.01.
Remember, in considering that probability, that there was a period
of more than a million years in which the two planets could have

collided. With the very eccentric and extensive orbits of the six initial major planets there were fifteen possible pairs of planets that could have collided. If the probability for each pair was 0.01 then the probability of *some* collision event was $0.01 \times 15 = 0.15$ — so not at all unlikely. The numbers are only illustrative. An actual calculation for a slightly different model showed that some major event was more likely than not, but for that calculation the definition of a major event included a close interaction in which one or other of the planets was thrown out of the Solar System.

So it is claimed that the Capture Theory is *plausible* — no more, no less. The degree of plausibility has to be left to the judgement of the reader and those scientists who can understand the theoretical basis of what has been described. However, even if it were agreed that, as related so far, the theory gave *perfect* explanations for the origin and evolution of the Solar System that would still not exclude the possibility that some new evidence could discredit it completely. Nor can the possibility be excluded that some other theory, based on a completely different scenario, could be advanced that would be equally, or even more, plausible. Having said that, experience with this subject stretching in the scientific era over more that 200 years, suggests that it is not easy to find good explanations for the wide variety of features of the Solar System. Any theory that appears to do so should not be lightly dismissed.

The main test to be applied to decide whether or not a theory should be taken seriously is to consider its consistency and coherence — or lack of it. Does the story flow with a continuity of events like the narration of a long novel — or is it a disjointed series of short stories without obvious links? Does it involve a minimum number of assumptions and events to explain the diverse features of the system? This last point is really a reiteration of Occam's razor, mentioned in Chapter 1. A theory that gave a series of *ad hoc* explanations for almost every feature of the Solar System, lacking any common root, would not inspire confidence. For example, in Chapter 40 we saw that many anomalous isotopic features have been assumed by meteoriticists to have come from various sources outside the Solar System. The planetary collision model, and the associated nuclear explosion — supporting evidence for which exists in other features of the Solar System as well — derives all the anomalies in one event.

The account given here divides into two parts. The first describes the origin of the Solar System from the initial interstellar medium through to the formation of a set of major planets with satellites. The second part is concerned with the evolution of the initial system. It starts with planets, accompanied by satellites, in highly elliptical but evolving orbits and ends with the Solar System in much the state we see it today. The two parts are intimately linked in that for the evolutionary processes to be able to take place it is necessary to have the starting point provided by the first part. There must be planets in highly elliptical orbits undergoing precession otherwise a collision cannot occur. It is from that one event, the collision, that so much else follows. Of course, if some other model can be found for the formation of planets that also led to major planets in similar highly eccentric orbits, then the second evolutionary part of this model could still apply.

The strength of the model is that sequences of events are causally related. As a consequence of a supernova there is cooling of the interstellar medium to give a cool dense collapsing cloud. Turbulence in the cloud gives the collisions of turbulent elements that lead to stars. The cluster of stars, falling inwards in a collapsing cloud, reaches a dense embedded stage where interactions between condensed stars and either protostars or dense compressed regions take place and planets are formed by the capture-theory process. The collapsing planets develop disks that provide material for satellite formation. The same kind of sequencing can be made for the second part that also has another feature — a single event, the planetary collision, which leads to explanations of features as different as the origin and characteristics of Triton and isotopic anomalies in meteorites.

It is interesting that, in some respects, observations have sometimes supported ideas that had previously been advanced for the Capture Theory and the planetary-collision hypothesis. When the idea of the capture process was first advanced the only planetary system known to exist was the Solar System so arguments that the probability of its formation was low were not relevant. The detection of exoplanets changed that situation; now probability arguments *were* relevant but the observation of the dense embedded stage of the evolution of a galactic cluster came to the aid of the Capture Theory. Interactions between compact stars — either as YSOs or in the main-sequence stage — would be so frequent that a large proportion of Sun-like stars should have an accompanying family of planets. Again, the original work by Holden and me in 1995 to explain isotopic anomalies in meteorites was based on the idea of an enhanced D/H ratio in an ice-silicate mixture within a planet but without much evidence that it could occur. The later observation of enhanced D/H ratios in icy grains in cool dense clouds and protostars made it extremely likely that early planets produced from protostar or compressed cool-dense-cloud material would have an icy shell with an enhanced D/H ratio.

The evolutionary pattern of development of the model is illustrated in Figure 41.1 in the form of a simple flow diagram with colour coding to show related features. The material in blue would relate to all planetary systems and the remainder, following the planetary collision, just to the Solar System. It should again be stressed that all the really critical features of the model, and some of the less critical features, have been subjected to detailed computer simulations. It should also be said that not everything that comes out of the capture-theory approach has been included — for example, ideas about lunar magnetism — but such excluded details are very peripheral to the main description.

Naturally, the theory may not be correct and eventually there may be a better one.

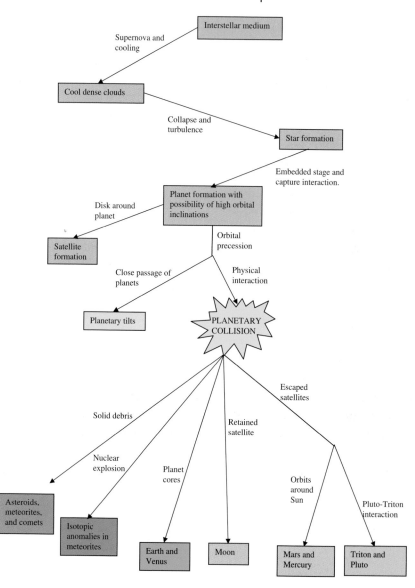

Figure 41.1 Development of the Solar System from a supernova to its present form.

Epilogue

An Autumn Evening

Greg settled himself down into the armchair with a contented sigh and gazed into the log fire. Lydia had excelled herself that evening; in particular the venison, his favourite treat, had been cooked to perfection. She was settling the kids into bed and she would soon bring in the coffee to round off what had been a perfect meal. Then they would both settle down to read — not talking, but still enjoying each other's company.

He glanced at the coffee table at his elbow on which was the journal that had just arrived and which contained the papers presented at the spring conference on *The Origin of the Solar System*. He had not presented a paper himself but Henry, his research student, had given a talk on '*The Detection of Interstellar Dust within the Solar System*' that had been very well received. It was an excellent meeting and, despite the many problems that remained, he had a sense that real progress was being made. He recalled that recent work on the top-down formation of planets and on the retrograde orbits of exoplanets had provoked lively debates but it was only by considering and discussing all possible ideas that a realistic model would eventually emerge. After all, everyone agreed on the basic model — that the Solar System, and planetary systems in general, had grown out of a nebula — and it was just the details of the formation of the various types of body that had to be worked out. Actually, he recalled, perhaps not everyone agreed on the nebula model. There was one participant who had raised a question about some alternative model involving a tidal interaction between stars — or something like that — but the Chairman had commented, in his rather acerbic way, that they were there to discuss serious science and could not spend valuable time discussing offbeat ideas. The Chairman also reminded the audience that tidal theories had been investigated and found wanting almost one hundred years before and there was little point in trying to revive them now. At the time it had occurred to Greg that the Chairman was being a little inconsistent here — after all, nebula ideas had been revived after being abandoned for nearly a century.

Lydia arrived with the coffee. Greg picked up the journal and began to read.

Bibliography

Chapter 1

Kuhn, Thomas S. (1970) *The Structure of Scientific Revolutions* (University of Chicago Press, Chicago).

Chapter 3

Robert Wilson (1997) *Astronomy Through the Ages* (Taylor & Francis, London).

Chapter 4

Hall, A. Rupert (1981) *From Galileo to Newton* (Dover Publications, New York).

Chapter 5

Chown, Marcus (2011) *The Solar System* (Faber & Faber, London).

Chapter 6

Chown, Marcus (2011) *The Solar System* (Faber & Faber, London).

Chapter 7

Elkins-Tanton, Linda T. (2010) *Asteroids, Meteorites and Comets* (*The Solar System*, Facts on File (J)).

Chapter 9

Laplace, P.S. de (1796) *Exposition du Système du Monde* (Imprimerie Cercle-Social, Paris).

Roche, E. (1873) Essai sur la constitutionet l'origine du système solaire, *Mem. Acad. Montpellier*, **8**, 235–234.

Brush, Stephen G. (1996) *Nebulous Earth*: *The Origin of the Solar System and the Core of the Earth from Laplace to Jeffreys* (Cambridge University Press, Cambridge).

Brush, Stephen G. (1996) *Fruitful Encounters*: *The Origin of the Solar System and of the Moon from Chamberlin to Apollo* (Cambridge University Press, Cambridge).

Chapter 10

Chamberlin, T.C. and Moulton, F.R. (1900) Certain recent attempts to test the nebula hypothesis, *Science* **12**, 201–208.

Moulton, F.R. (1907) On the probability of the near approach of two suns and the orbits of material ejected from them under the stimulus of their mutual tidal disturbances, *Carn. Inst. Yb.* **5**, 168–169.

Nõlke, F. (1908) *Das Problem der Entwicklung unseres Planeten-systems. Aufstellung eine neuen Theorie nach vorgehende Kritik der Theorien von Kant, Laplace, Poincaré, Moulton, Arrhenius, u.a.* (Springer, Berlin).

Brush, Stephen G. (1996) *Nebulous Earth*: *The Origin of the Solar System and the Core of the Earth from Laplace to Jeffreys* (Cambridge University Press, Cambridge).

Brush, Stephen G. (1996) *Fruitful Encounters*: *The Origin of the Solar System and of the Moon from Chamberlin to Apollo* (Cambridge University Press, Cambridge).

Chapter 11

Jeans, J.H. (1917) The motions of tidally distorted masses with special reference to theories of cosmogony, *Mem. Roy. Astron. Soc.* **652**, 1–48.

Jeans, J.H. (1938) The origin of planets, *Science and Culture* **4**, 73–75.

Jeans, J.H. (1942) Origin of the Solar System, *Nature* **149**, 695.

Jeffreys, H. (1929) Collision and the origin of rotation in the solar system, *M. N. Roy. Astron. Soc. Supplement* **89**, 636–641.

Russell, H.N. (1935) *The Solar System and its Origin* (Macmillan, New York).

Spitzer, L. (1939) The dissipation of planetary filaments, *Astrophys. J.* **90**, 675–688.

Brush, Stephen G. (1996) *Nebulous Earth: The Origin of the Solar System and the Core of the Earth from Laplace to Jeffreys* (Cambridge University Press, Cambridge).

Brush, Stephen G. (1996) *Fruitful Encounters: The Origin of the Solar System and of the Moon from Chamberlin to Apollo* (Cambridge University Press, Cambridge).

Chapter 12

Aust, C. and Woolfson, M.M. (1973) On the accretion mechanism for the formation of a protoplanetary disk, *M. N. Roy. Astron. Soc.* **161**, 7–13.

Bondi, H. and Hoyle, F. (1944) On the accretion of matter by stars, *M. N. Roy. Astron. Soc.* **104**, 273–282.

Lyttleton, R. (1960) Dynamical calculations relating to the origin of the solar system, *M. N. Roy. Astron. Soc.* **121**, 551–569.

Schmidt, O.J. (1944) A meteoric theory of the origin of the Earth and planets, *Comptes Rendus (Doklady) Academie des Sciences de l'URSS* **45**, 229–233.

Brush, Stephen G. (1996) *Nebulous Earth: The Origin of the Solar System and the Core of the Earth from Laplace to Jeffreys* (Cambridge University Press, Cambridge).

Brush, Stephen G. (1996) *Fruitful Encounters: The Origin of the Solar System and of the Moon from Chamberlin to Apollo* (Cambridge University Press, Cambridge).

Chapter 13

Jeffreys, H. (1952) The origin of the solar system, *Proc. Roy. Soc.* **A214**, 281–291.

Weizsacher, C.H. von (1944) Über die Enstehung des Planetensystems, *Z. fur Astrophysik* **22**, 319–355.

Brush, Stephen G. (1996) *Nebulous Earth: The Origin of the Solar System and the Core of the Earth from Laplace to Jeffreys* (Cambridge University Press, Cambridge).

Brush, Stephen G. (1996) *Fruitful Encounters: The Origin of the Solar System and of the Moon from Chamberlin to Apollo* (Cambridge University Press, Cambridge).

Chapter 14

McCrea, W.H. (1960) The origin of the solar system, *Proc. Roy. Soc.* **A256**, 245–266.

McCrea, W.H. (1988) Formation of the solar system. Brief review and revised protoplanet theory, in *The Physics of Planets: Their Origin Evolution and Structure*, ed. S.K. Runcorn (Wiley, Chichester).

Brush, Stephen G. (1996) *Nebulous Earth: The Origin of the Solar System and the Core of the Earth from Laplace to Jeffreys* (Cambridge University Press, Cambridge).

Brush, Stephen G. (1996) *Fruitful Encounters: The Origin of the Solar System and of the Moon from Chamberlin to Apollo* (Cambridge University Press, Cambridge).

Chapter 16

Balli, J., Testi, L., Sargent, A and Carlstrom, J. (1998) Lifetimes of externally illuminated young stellar objects embedded in the Orion Nebula, *Astron. J.* **116**, 854–859.

Haisch, K.E., Lada, E.A, and Lada, C.J. (2001) Disk frequencies and lifetimes in young clusters, *Astrophys. J.* **553**, L153–L156.

Chapter 17

Butler, R.P. and Marcy, G.W. (1996) A planet orbiting 47 Ursae Majoris, *Astrophys. J.* **464**, L153–156.

Cameron, A.C. (2001) Extrasolar planets, *Physics World.*

Chapter 18

Brush, Stephen G. (1996) *Fruitful Encounters: The Origin of the Solar System and of the Moon from Chamberlin to Apollo* (Cambridge University Press, Cambridge).

Chapter 19

Armitage, P.J. and Clarke, C.J. (1996) Magnetic braking of T-Tauri stars, *M. N. Roy. Astr. Soc.* **280**, 458–468.

Hoyle, F. (1960) On the origin of the solar nebula, *Q. J. R. Astr. Soc.* **1**, 28–55.

Lynden-Bell, D. and Pringle, J.E. (1974) The evolution of viscous disks and the origin of nebula variables, *M. N. Roy. Astr. Soc.* **168**, 603–637.

Chapter 20

Cameron, A.G.W. (1978) The primitive solar accretion disk and the formation of planets, *The Origin of the Solar System.* Ed. S.F. Dermott (Wiley, Chichester).

Chapter 21

Blum, J., Wurm, G., Kempf, S. *et al.* (2000) On growth and form of planetary seedlings, *Phys. Rev. Lett.* **85**, 2426–2429.

Goldreich, P. and Ward, W.R. (1973) The formation of planetesimals, *Astrophys. J.* **183**, 1051–1061.

Konacki, M. (2005) An extrasolar giant planet in a close triple-star system, *Nature*, **436**, 230–233.

Rogers, T.M., Lin, D.N.C. and Lau, H.H.B. (2012) Internal Gravity Waves Modulate the Apparent Misalignment of Exoplanets around Hot Stars, *Astrophys. J. Lett.* **758**, Issue 1.

Safronov, V.S. (1972) *Evolution of the Protoplanetary Cloud and Formation of the Earth and Planets* (Israel Program for Scientific Translations, Jerusalem).

Weidenschilling, S.J., Donn, B. and Meakin, P (1989) *The Formation and Evolution of Planetary Systems*, eds. H.A. Weaver and I. Danley (Cambridge University Press, Cambridge).

Chapter 22

Wetherill, G.W. and Stewart, G.R. (1989) Accumulation of a swarm of small planetesimals, *Icarus* **77**, 330–357.

Wetherill, G.W. and Stewart, G.R. (1993) Formation of planetary embryos: Effects of fragmentation, low relative velocity and independent variation of eccentricity and inclination, *Icarus* **100**, 190–209.

Chapter 23

Cole, G.H.A. and Woolfson, M.M. (2013) *Planetary Science: The Science of Planets around Stars*. Second Edition (Taylor & Francis, London).

Kokubo, E and Ida, S. (1998) Oligarchic growth of planetesimals, *Icarus* **131**, 171–178.

Ward, W.R. (1997) In ASP Conf. Ser. Eds. Hettig, T. and Hahn, J.M., *On the formation and migration of protoplanets*, Proc. 1ˢᵗ Int. Origins Conf. Astron. Soc. Pac. (San Francisco).

Chapter 24

Beckwith, S.V.W. and Sargent, A. (1996) Circumstellar disks and the search for neighbouring planetary systems, *Nature*, **393**, 139–144.

Boss, A.P. (2000) Possible rapid gas giant formation in the solar nebula and other protoplanetary disks, *Astrophys. J.* **536**, L101–L104.

Boss, A.P. (2008) Flux-limited diffusion approximation models of giant planet formation by disk instability, *Astrophys. J.* **677**, 607–615.

Chapter 26

Golanski, Y. and Woolfson, M.M. (2001) A smoothed particle hydrodynamics simulation of the collapse of the interstellar medium, *M. N. Roy. Astron. Soc.* **320**, 1–11.

Goldsmith, D.W., Habig, H. and Field, G.B. (1969) Thermal properties of interstellar gas heated by cosmic rays, *Astrophys. J.* **158**, 173–184.

Hayashi, C. (1966) Evolution of protostars, *Ann. Rev. Astron. Astrophys.* **4**, 171–192.

Chapter 27

Bonnell, I.A., Clarke, C.J., Bate, M.R. and Pringle, J.E. (2001) Accretion in stellar clusters and the initial mass function, *M. N. Roy. Astron. Soc.* **324**, 573–579.

Krumholtz, M.R., McKee, C.F. and Klein, R.I. (2005) The formation of stars by gravitational collapse rather than competitive accretion. *Nature* **438**, 332–334.

Woolfson, M.M. (1979) Star formation in a galactic cluster, *Phil. Trans. R. Soc. Lond.* **A291**, 219–252.

Chapter 28

Kroupa, P. (1995) Star cluster evolution, dynamical age estimation and the kinematic signature of star formation, *M. N. Roy. Astron. Soc.* **277**, 1522–1540.

Lada, C.J. and Lada, E.A. (2003) Embedded clusters in molecular clouds, *Ann. Rev. Astron. Astrophys.* **41**, 57–115.

Chapter 29

Oxley, S. and Woolfson, M.M. (2003) Smoothed particle hydrodynamics with radiation transfer, *M. N. Roy. Astron. Soc.* **343**, 900–912.

Oxley, S. and Woolfson, M.M. (2004) The formation of planetary systems, *M. N. Roy. Astron. Soc.* **348**, 1135–1149.

Woolfson, M.M. (1964) A capture theory of the origin of the solar system, *Proc. R. Soc. Lond.* **A282**, 485–507.

Chapter 30

D'Antona, F. and Mazzitelli, I. (1998) http//www.inporzio.astro.it/–dantona/prems.html.

Melita, M.D. and Woolfson, M.M. (1996) Planetary commensurabilities driven by accretion and dynamical friction, *M. N. Roy. Astron. Soc.* 280, 854–862.

Woolfson, M.M. (2003) The evolution of eccentric planetary orbits, *M. N. Roy. Astron. Soc.* **340**, 43–51.

Chapter 31

Woolfson, M.M. (2004) The stability of evolving planetary orbits in an embedded cluster, *M. N. Roy. Astron. Soc.* **348**, 1150–1156.

Woolfson, M.M. (2011) *On the Origin of* Planets: *By Means of Natural Simple Processes*, pp. 119–130 (Imperial College Press, London).

Chapter 32

Kimura, H., Mann, I. and Jessberger, E.K. (2003) Composition, structure and size distribution of dust in the local interstellar cloud, *Astrophys. J.* **583**, 314–321.

Schofield, N. and Woolfson, M.M. (1982) The early evolution of Jupiter in the absence of solar tidal forces, *M. N. Roy. Astron. Soc.* **198**, 947–959.

Vicento, S. and Alvez, J. (2005) Size distribution of circumstellar disks in the Trapezium Cluster, *Astron & Astrophys.* **441**, 195–205.

Woolfson, M.M. (2004) The formation of regular satellites, *M. N. Roy. Astron. Soc.* **384**, 419–426.

Chapter 33

Dormand, J.R. and Woolfson, M.M. (1977) Interactions in the early solar system, *M. N. Roy. Astron. Soc.* **180**, 243–279.

Hirano T., Narita N., Shporer A., Sato B., Aoki W. & Tamura M. (2011) A possible tilted orbit of the super-Neptune HAT-P-11b, *Publ. Astron. Soc. Japan*, **63**, 531–536.

Kozai, Y. (1962) Secular perturbations of asteroids with high inclination and eccentricity, *Astrophys. J.* **67**, 591–598.

Parise, B., Caux, E., Castets, A., *et al.* (2005) HDO abundance in the envelope of the solar type protostar IRAS 16293-2422, *Astron & Astrophys.* **431**, 547–554.

Parise, B., Ceccarelli, C., Tielens, A.G.G.M. *et al.* (2002) Detection of double deuterated methanol in the solar type protostar IRAS 16293-2422, *Astron & Astrophys.* **393**, L49–L53.

Roberts, H., Herbst, E. and Millar, T.J. (2003) Enhanced deuterium fractionation in dense interstellar cores resulting from multiply deuterated H_3^+, *Astrophys. J.* **591**, L41–L44.

Chapter 34

Hutchison, R. (1983) *A Search for our Beginning* (Oxford University Press, Oxford).

Chapter 35

Dormand, J.R. and Woolfson, M.M. (1977) Interactions in the early solar system, *M. N. Roy. Astron. Soc.* **180**, 243–279.

Woolfson, M.M. (2011) *On the Origin of* Planets: *By Means of Natural Simple Processes*, pp. 190–201 (Imperial College Press, London).

Chapter 36

Dormand, J.R. and Woolfson, M.M. (1977) Interactions in the early solar system, *M. N. Roy. Astron. Soc.* **180**, 243–279.

Woolfson, M.M. (2011) *On the Origin of* Planets: *By Means of Natural Simple Processes*, pp. 203–218 and 402–403 (Imperial College Press, London).

Chapter 37

Connell, A.J. and Woolfson, M.M. (1983) Evolution of the surface of Mars, *M. N. Roy. Astron. Soc.* **204**, 1221–1240.

Lamy, P.L. and Burns, J.A. (1972) Geometrical approach to torque free motion of a rigid body having internal energy dissipation, *Am. J. Phys.* **40**, 441–444.

Woolfson, M.M. (2000) *The Origin and Evolution of the Solar System* (Institute of Physics, Bristol).

Woolfson, M.M. (2011) *On the Origin of* Planets: *By Means of Natural Simple Processes*, pp. 426–428 (Imperial College Press, London).

Chapter 38

Benz, W., Slattery, W.L. and Cameron, A.G.W. (1986) The origin of the Moon and the single impact hypothesis, *Icarus* **66**, 515–535.

McCord, T.B. (1966) Dynamical evolution of the Neptunian system, *Astron. J.* **71**, 585–590.

Woolfson, M.M. (1999) The Neptune-Triton-Pluto system, *M. N. Roy. Astron. Soc.* **304**, 195–198.

Chapter 39

Bailey, M.E. (1983) The structure and evolution of the solar system comet cloud, *M. N. Roy. Astron. Soc.* **204**, 603–633.

Bailey, M.E., Clube, S.V.M. and Napier, W.M. (1990) *The Origin of Comets* (Butterworth Heinemann, Oxford).

Napier, W.M. and Dodd, R.J. (1973) The missing planet, *Nature* **242**, 250–251.

Chapter 40

Holden, P. and Woolfson, M.M. (1995) A theory of local formation of isotopic anomalies in meteorites, *Earth, Moon and Planets* **69**, 301–236.

Woolfson, M.M. (2011) *On the Origin of* Planets: *By Means of Natural Simple Processes*, pp. 269–293 and 435–443 (Imperial College Press, London).

Chapter 41

Kuhn, Thomas S. (1970) *The Structure of Scientific Revolutions* (University of Chicago Press, Chicago).

Index